UNDERSTANDING THE

IGNITION SYSTEM

A straightforward guide to help you understand how to get the best performan ost systems i

© GUNSON LIMITED 1987. Gunson Limited, Pudding Mill Lane, London E15 2PJ, England.
All rights reserved. No part of this publication may be reproduced, stored in a retrieval system, or transmitted in any form or by any means, electronic, mechanical, photocopying, recording or otherwise, without the prior permission of the publishers.

CONTENTS

INTRODUCTION

As any Boy Scout will tell you, in order to start a fire to cook a camping breakfast, you basically need three things. First a suitable place for the fire, secondly a suitable fuel and finally some means of igniting it. If any one of these key elements is missing or unsatisfactory, then there will either be no fire, or it won't burn very well.

Spark ignition engines need very similar factors to achieve good combustion, which is a fundamental requirement for maximum performance and economy.

In these engines the fire burns in an enclosed cylinder, the actual area being known as the 'combustion chamber'. The fuel (petrol), mixed with air (all fires need oxygen) is drawn into the cylinder and compressed where a good 'fat' spark is needed at exactly the right time, for it to burn properly. The word 'fat' is used to denote a spark of sufficient strength or ignition energy, for its heat to ignite the fuel.

As with the Boy Scouts fire if any one of those three basic factors is unsatisfactory, then combustion will be affected. In a well designed and maintained engine this will not be the case but, there should be an excess of spark energy in the ignition system, so as to ignite the fuel mixture under unfavourable conditions such as when the engine is cold.

Furthermore if the Boy Scout lights his fire late, the whole days working schedule could be upset. If the spark is late in an engine, the whole working cycle is affected and there will be a loss of power, overheating and increased exhaust pollution. Much the same will apply if the spark is early, but whereas the burn-time of the air/fuel mixture in the engine is more or less constant, the time allowed for it decreases as the engine goes faster. In order then for the fuel to burn completely at higher engine speeds the spark has to be provided earlier, in other words the ignition has to be advanced.

Another function of the ignition is, therefore to regulate the precise moment when the spark occurs, so that it starts the combustion process at exactly the right time in each cycle of operations and under all engine conditions.

In order for this to happen the actual ignition process begins long before the spark ignites the fuel in the cylinder. This process in a conventional ignition system comprises the stages opposite.

This process is a remarkable one, not only because of its intricacy but because it occurs at between 20,000 and 30,000 times a minute in the system of an average motor car and at more than 40,000 times a minute in electronic systems used on some motorcycle and other high speed engines.

It can be illustrated in a classic graph which shows how the battery current, primary voltage and secondary voltage relate to one another over a period of time measured in milliseconds (one thousandth part of a second), often abbreviated to ms.

1.	the supply of electricity (Battery Current)	by the battery
2.	the switching of this supply into separate phases (Primary Voltage)	by the contact breaker points
3.	the passage of these phases	by the primary winding of the coil
4.	the resultant generating of very much more powerful pulses of electricity (Secondary Voltage)	by the secondary winding of the coil
5.	the distribution and timing (advanced or retarded) of these pulses to the cylinders.	by the distributor
6.	the production of sparks to ignite the mixture	by the sparking plug

MAGNETO IGNITION

A magneto can be considered as a generator, coil and distributor in one unit and, as such does not require a battery.

Although there are different types of magneto, in general they are all more expensive to produce than coil systems, especially for multi-cylinder engines and produce their poorest spark at low (cranking) speeds.

For these reasons their use is generally restricted to small single cylinder engines such as those used in small motor-cycles, lawn mowers, generators and the like.

In the version shown here four permanent magnets are located in the flywheel. As this rotates the magnets generate a current in the primary winding (W1). When the primary circuit is interrupted by the contact breaker opening, a high tension pulse is created in the secondary winding (W2), which produces a spark at the plug.

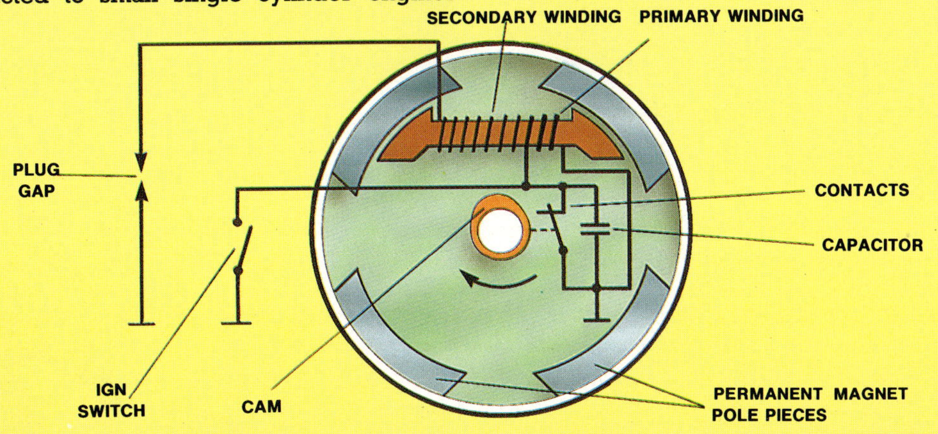

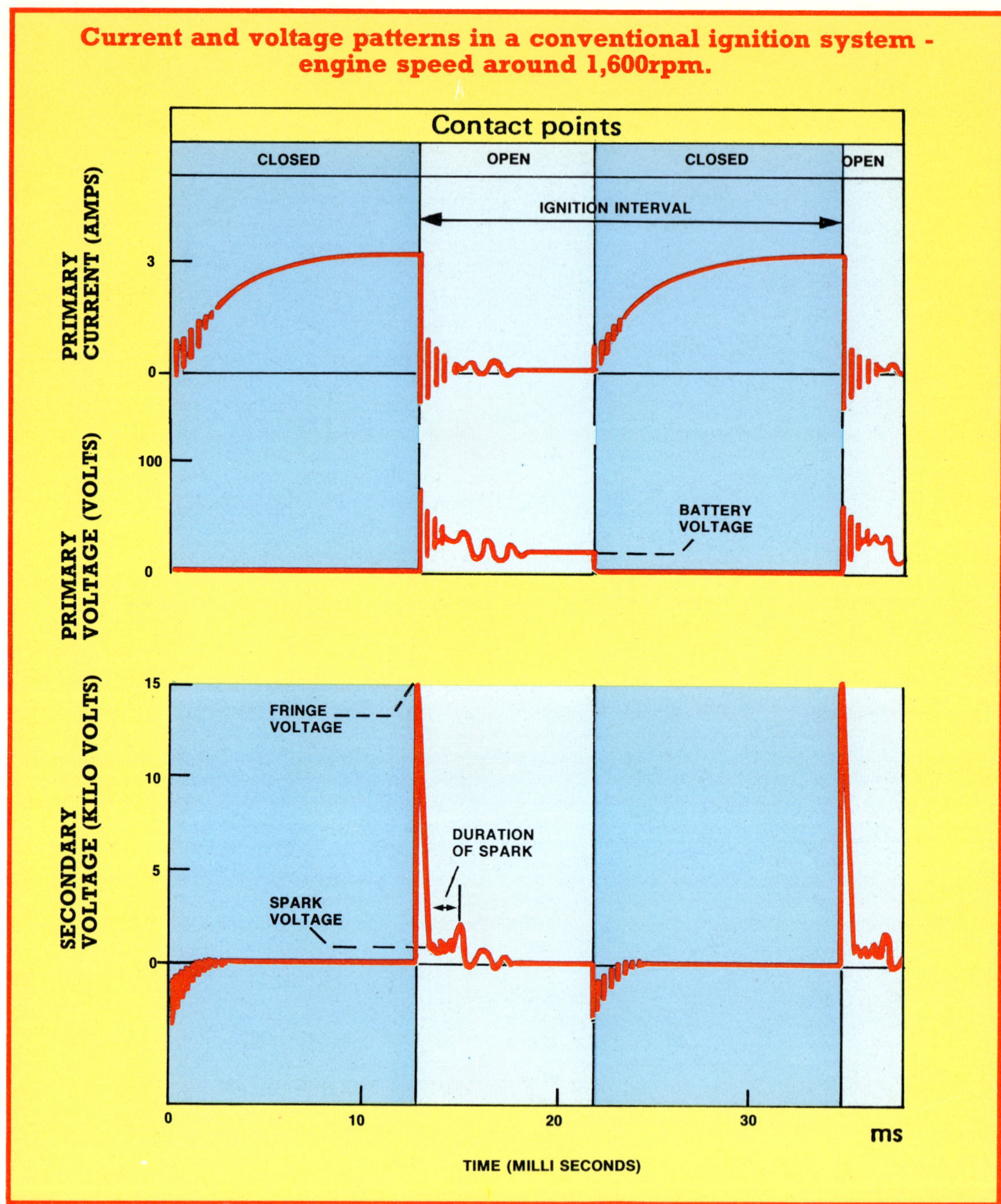

By far the majority of professional tuning agencies now use an instrument called a oscilloscope for checking out a vehicles ignition system. With this the operator can determine what, if any, ignition faults exist from a waveform displayed on the oscilloscope screen.

In addition to this however, the waveforms can also give some idea of what happens in the ignition system and over what time scale. The patterns shown here are from a typical four cylinder engine running at around 1,600rpm and are, in fact, three different displays – the primary current, the primary voltage and the secondary voltage – shown together in one chart to illustrate the relationship of one to the others.

Although the chart refers to the opening and closing of the contact breakers, virtually the same sequence of events takes place in any inductive electronic system, except that the primary current is switched by a transistor.

PRIMARY CURRENT

With the ignition switched on and the contact points closed, it takes from around 10 to 15 milli-seconds for the primary circuit to build up to its full valve. This is because a magnetic field is being developed in the primary winding of the coil which induces a voltage into the winding which opposes battery voltage.

This is the dwell period and obviously decreases in time as the engine speed increases. In some cases this could result in the points opening before the primary current has reached its maximum value. Should this happen the secondary voltage may not reach its firing level and a high speed misfire would result.

In a conventional (contact breaker) ignition system the maximum current would be in the region of 4 amps.

PRIMARY VOLTAGE

When the contact points are closed, the primary winding is in effect earthed so the voltage reading is zero. However once the points open, the magnetic field in the coil collapses and induces a voltage into the primary winding. The value of this induced voltage is dependent upon a number of factors, including the rate of collapse and therefore the effectiveness of the capacitor.

This induced voltage quickly dies out in a number of decreasing oscillations, losing energy in the form of heat which is dissipated throughout the circuit. This also accounts for the pulses shown in the primary current scale when the points open.

Although the induced voltage shown here is in the region of 80 volts, it can exceed 300 volts.

SECONDARY VOLTAGE

As the magnetic field collapses when the contact points open its lines of force are cut by the turns of wire in both the primary and secondary windings. However as there are many more turns of wire in the secondary, the induced voltage in that winding will be greater.

In order to initially bridge the gap, at the plug the secondary voltage has to reach the 'firing' level of around 15,000 volts. However once having made the gap conductive, the spark can be maintained at a much lower level. This is shown as the spark voltage. Once the voltage drops below this level, the spark is extinguished and what voltage remains dies out in much the same way as the induced voltage in the primary winding.

Even in a relatively low performance vehicle such as the average family saloon car, the ignition system is put under considerable pressure to give good performance under all conditions. It therefore needs to be of a high quality to begin with, but also needs to be regularly and well maintained.

Admittedly modern systems, particularly the various electronic versions, have greatly extended service intervals - in some cases the only listed service requirement may be to change the sparking plugs every 12,000 miles. In many ways this can be a retrograde step for many motorists with these systems fitted to their cars tend to carry out even less maintenance and fewer checks. Consequently the system can become dirty, terminals can become corroded and tracking can take place.

This would appear to be borne out by various breakdown and tuning surveys, where more faults are invariably recorded on ignition systems than in any other single field. Strangely enough a considerable number fitted with electronic ignition appear to have timing errors and, according to the AA most breakdowns involve cars less than five years old.

Nevertheless ignition systems have improved to give:

- **better secondary voltage and spark energy, even at high engine speeds.**
- **greater ability to overcome shorts or shunts due to damp or deposits on the plugs.**
- **longer service life of parts, especially contact breakers and in spite of any possible drawbacks.**
- **lower maintenance (and therefore service costs).**

A SUMMARY OF IGNITION SYSTEMS:

IGNITION SYSTEM TYPE	ADVANTAGES	DISADVANTAGES
INDUCTIVE Conventional coil/contact-breaker ignition.	**Cheaper parts. Easy to fit. Easy to maintain but needs accurate contact-breaker point setting.**	**Prone to misfiring at very high speeds. Prone to loss of secondary voltage (due to dirt, damp etc). Require regular maintenance (at least every 6,000 miles)**
INDUCTIVE SEMI-CONDUCTOR Often referred to as TAC (Transistor Assisted Contact). A transistor (solid state relay) is used to switch the primary current – the contact breakers being retained only to trigger the transistor.	**Fairly cheap parts. Can be used to give higher secondary voltage. Reduced maintenance. Easy to convert back to conventional in the event of electronic failure.**	**Still prone to misfire at high speed to some extent. Not so much of an improvement on 6 or 8 cylinder engines. Still prone to losses of secondary voltage (as in conventional systems).**
BREAKERLESS SEMI-CONDUCTOR Still uses a transistor to switch the primary current, but employs different types of triggering device such as – pulse generator, optical switch or Hall Effect trigger. The contact points are eliminated.	**As with previous system but with better misfire performance up to around 10,000rpm (four cylinder engines). Less likely to misfire at high speed and better for 6 or 8 cylinder engines.**	**More expensive. Can be difficult to convert back to conventional and impossible if factory fitted system.**
CAPACITOR DISCHARGE Often referred to as CD systems, these use a transformer to charge a capacitor to around 300 volts and which when switched by a thyristor discharges through the primary winding of a coil. This generates an extremely high voltage pulse in the secondary winding and produces a very strong but short lived spark at the plug.	**Very high powered spark produced, but of very short duration – many systems incorporate electronic spark extenders to overcome this. Rapid and high rise of secondary voltage means that it can punch through shunts, such as fouled plugs, which in other systems would cause a misfire. Low secondary voltage loss.**	**Comparitively expensive. Spark duration may be too short for badly designed engines or those designed to run on weak fuel mixtures. Any follow on extended sparks may be weak with insufficient energy to ignite the fuel.**

The conventional (contact-breaker) ignition system then has considerable disadvantages compared with most electronic systems and is far less reliable. Having said that, a conventional system in good condition is perfectly adequate for the average family car and these disadvantages can be minimised by regular and accurate maintenance.

The Achilles heel of the conventional system is the contact-breaker for it not only affects the condition of the spark but also its timing. For this reason it is vitally important to make sure the points are correctly set before checking the ignition timing.

This book aims to clarify how ignition systems work, what the individual components do, how to maintain them and what to do if they should go wrong.

BASIC ELECTRICS

To many people, the electrical system of a car or motorcycle is made up of a battery, a starter motor, a generator and a number of lamps, all connected up by a complicated jumble of different coloured cables and with strange little black boxes dotted around, seemingly at random. How it all works is a complete mystery.

Although the whole layout may appear terribly complex, in reality it is made up of a number of individual circuits, each of which is fairly simple and can usually be traced quite easily, because of those different coloured cables.

Basically an electrical circuit is a distribution network, something like a domestic plumbing system and probably the easiest way to understand how it works is to relate it to a simple water layout.

In the hydraulic circuit a pump draws water from a collector tray and delivers it to a tank. From there it flows through a number of taps (when they are open) back to the collector tray.

The pressure in the circuit is maintained by the head of water in the tank which, in turn is dependant upon how much the pump can deliver and how many (and what size) taps are in use.

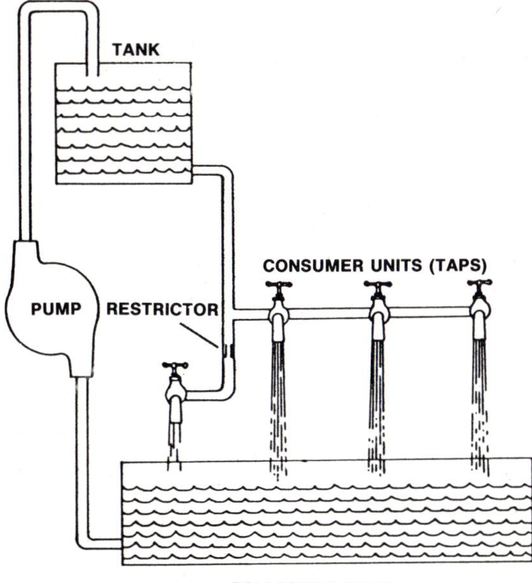

The volume of water passing through the circuit will also depend upon the size and number of taps in use and the pressure supplied from the tank. In addition any restrictor in the pipe can reduce the flow and also the pressure on the tap side. The taps themselves can cause a restriction, particularly if they are not fully open.

All three factors, the rate of flow, the pressure and the restrictions are then related, change any one and it must affect one or both of the others. Furthermore the volume of water flowing through the circuit is dependant upon all three factors.

Compare this with a simple vehicle electrical circuit. Think of the generator as a pump, the battery as a (storage) tank and consumer items such as the lamps, windscreen wiper and starter motor, together with their switches as the taps.

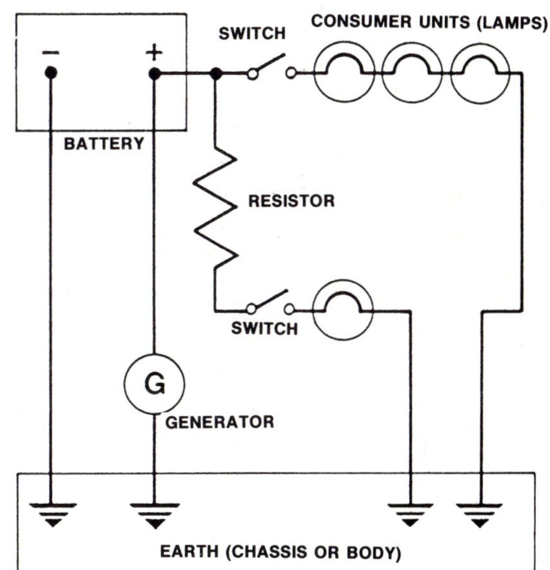

Switch on any of the consumer units and you will get a flow of current, measured in amps and forced through by the pressure – volts. Any restriction to flow is a resistance, measure in ohms.

As in the water circuit, all of these are related to one another, change any one and it will affect either one or both of the others. In an electrical circuit, the relationship of one factor to the others can be calculated by the formula known as Ohms Law where:

$$\text{Amps} = \frac{\text{Volts}}{\text{Ohms}}$$

Transposed this could be either:

$$\text{Ohms} = \frac{\text{Volts}}{\text{Amps}}$$

or

$$\text{Volts} = \text{Amps} \times \text{Ohms}$$

A further value often quoted is that of power consumed. This more or less equates to the volume of water passing through the circuit and is measured in Watts and is derived from the formula Watts = Volts × Amps. This can be transposed to show that a 36 watt bulb, used in a 12 volt circuit, would consume 3 amps.

There are basically two kinds of circuit in either the hydraulic or electrical applications, there are series and parallel circuits. In a series circuit the consumer items follow on one after the other and there will be a pressure drop across each. This means that the unit at the end of the line will deliver considerably less than the first one. In a parallel circuit on the other hand, each consumer unit is on a a separate line , the pressure at each will be more or less the same and provided there are no restriction in the lines each will deliver the same. Most vehicle circuits are of this (parallel) type.

In both applications (hydraulic and electric), it is fairly obvious that to create any flow (water or current) the circuit must be complete. A break anywhere in the pressure side of a hydraulic line would allow the water to run out, equating to a short circuit in electrical terms. A blockage, on the other hand, would stop any flow of water – electrically this would be the equivalent of an open circuit, a disconnected cable or such.

Finally, the collector tray shown in the hydraulic circuit doesn't strictly equate to anything in an electrical circuit, but can be considered as the chassis, frame or body, providing an earth return to the battery.

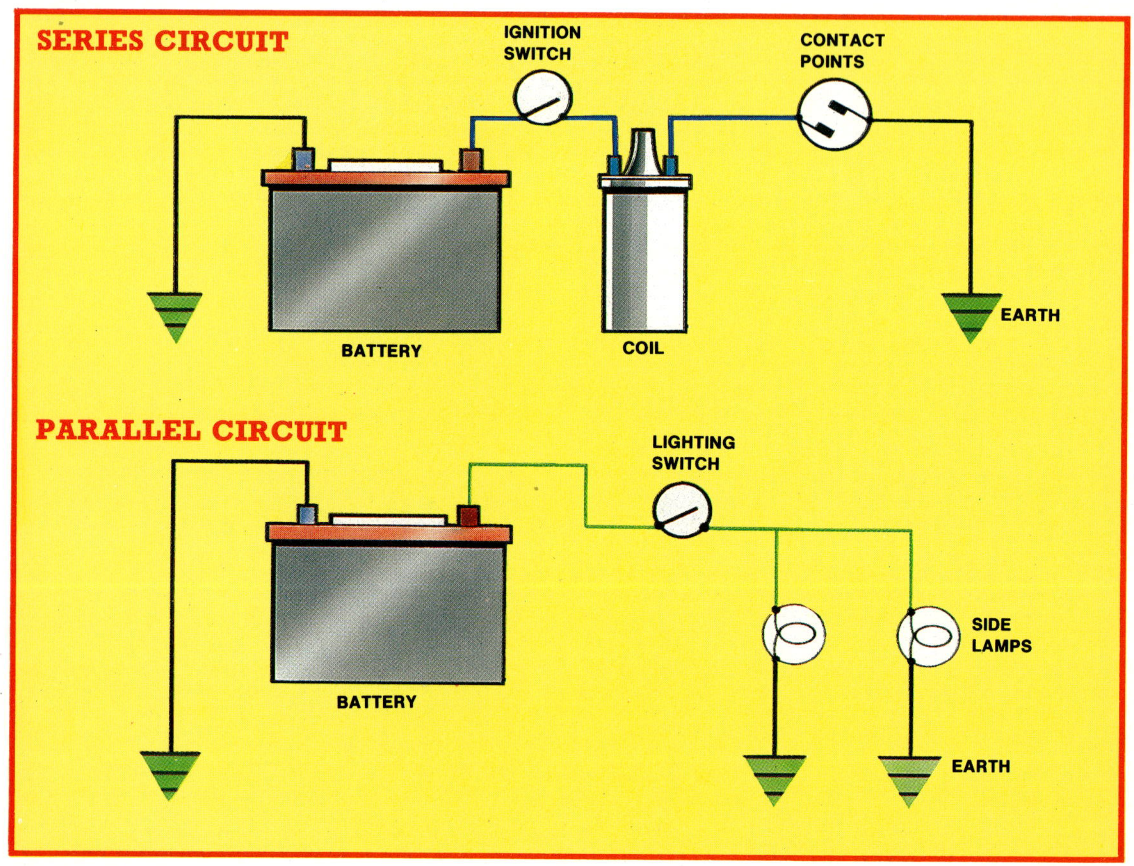

TOOLS AND EQUIPMENT

The tools and equipment required for working on the ignition system depend to some extent on the vehicle in question and to some extent on its age. There are however some tools common to all.

First among these will be a spanner for removing and refitting the plugs. This is generally either a form of box spanner or extra length socket, but as there are different sized plugs, the spanners too come in different sizes, so if buying one make sure it's the right one. Special spanners with a swivel joint incorporated into the handle can be especially useful on some engines.

A set of feeler gauges will be required to check the size of the gap between the electrodes of the sparking plug, these can be the same as those used for checking the contact breaker gap. Feeler gauges are available in both Imperial (inch) sizes and metric, most vehicle manufacturers now give the settings in metric sizes, but this wasn't the case a few years ago - check in the handbook.

Checking the contact breaker gap is one area where age comes into the equation for on a number of cars built before the late sixties, the setting may only be given as a fixed measurement of say 0.025inches, whereas cars built in the mid seventies and later only a dwell angle or period may be given. Some manufacturers give both.

Checking the setting when only the dwell is given, and for a more accurate setting when both are, requires the use of a dwell

A selection of useful hand tools for working on the Ignition.

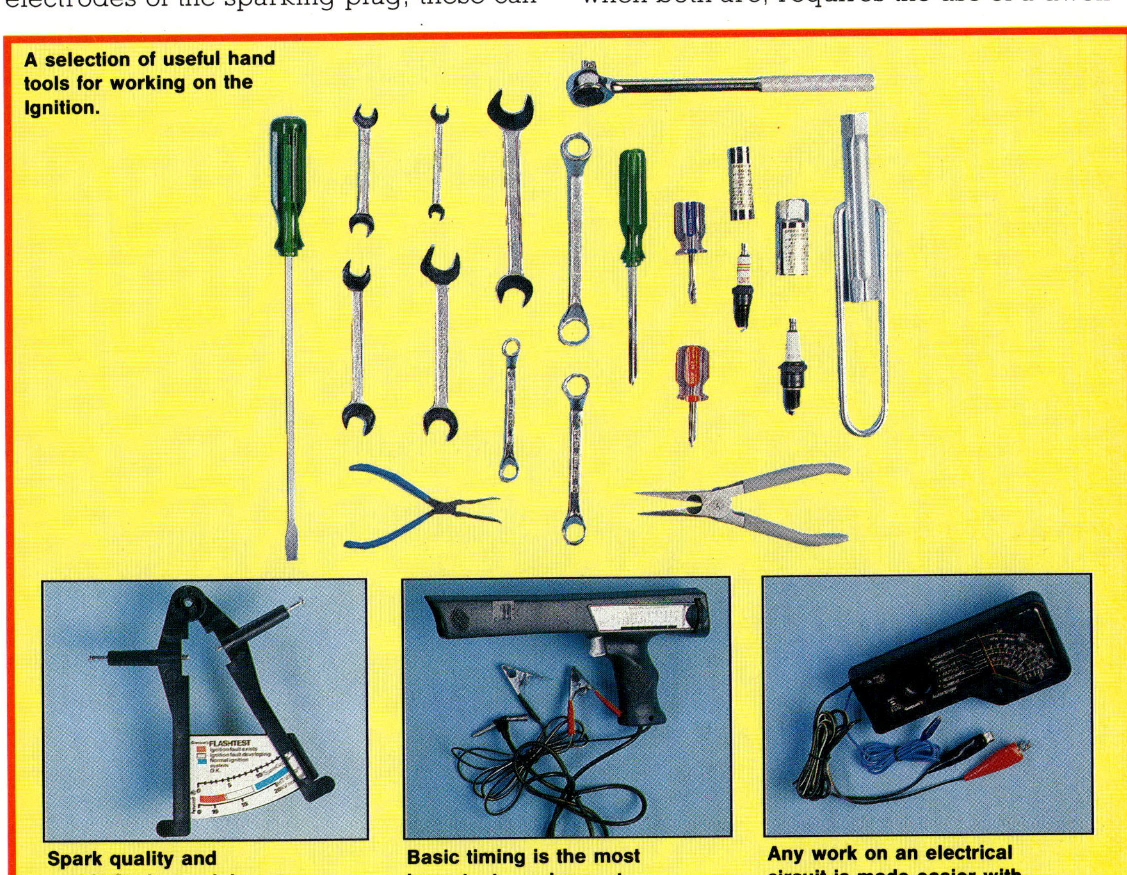

Spark quality and regularity is crucial – Flashtest easily checks this.

Basic timing is the most important requirement – Tachostobe is a good timing light.

Any work on an electrical circuit is made easier with a good test meter like Autoranger.

meter. These generally come as one mode of a multi-meter instrument, many of which will also have facilities for measuring volts, amps, ohms and engine rpm (tachometer) - if buying one of these, take into account the clarity of the scale - there's not much point in having a meter of any sort if you can't readily make out what the reading is.

If the major reason for all these tests is to make sure the end product (the spark) is sufficiently strong to ignite the mixture, then checking the value of the spark could eliminate these other checks - this can be done using the Flashtest.

Much of what applies to the contact breaker point settings also applies to checking that the ignition is correctly timed. In the old days this was generally done with the engine static, which was all very well but made no allowances for wear and play in the camshaft to distributor drive train, which could easily make a difference of 3 or 4 degrees - unacceptable by modern standards.

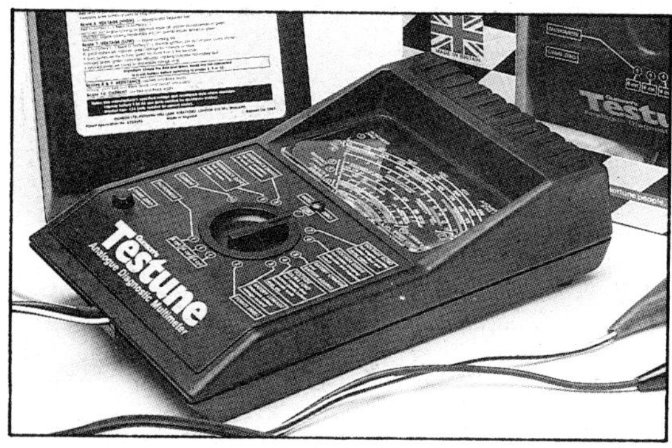

Nowadays, generally only a dynamic setting is given, that is one with the engine running and which means that it can only be checked with a stroboscopic light (strobe) that 'flashes' in unison with the spark, usually at number one cylinder.

Basically there are two kinds of strobe available, those with a neon tube and others with a xenon (pronounced zenon) tube. The neon versions are cheaper, but have a poorer light output and in some cases a longer 'flash' which makes the timing mark appear as a stripe. The xenon types give a far better light but need an outside power source, this can be from batteries enclosed within the strobe itself, from the vehicle battery or from the mains (240 volt).

Dynamic timing is always quoted in the handbook at a set engine speed, a typical example of which would be 8 degrees BTDC (Before Top Dead Centre) at 850rpm. Unless a rev-counter (tachometer) is permanently fitted this means that one would normally be needed for this operation, however some strobes incorporate a form of engine rpm indicator.

Setting the timing at this one speed is no guarantee that it will be correct at other (higher) speeds, when the automatic advance mechanisms are functioning. For instance at around 4,000 (engine) rpm the spark could well be timed at 28°BTDC. However this could, among other things, be affected by the condition of the centrifugal advance weight springs.

Very few vehicle manufacturers provide a timing mark, with which this can be checked. This could be overcome by physically measuring the circumference of the flywheel or pulley concerned and working out where the mark should be. In some cases this would be extremely difficult and, in any case is open to error, however there are strobes available which can check this degree of advance automatically from the one marker, by delaying the flash. If buying one of these make sure it is speed related - some have a pre-set delay, which would only be correct at one engine speed and which may not be the one quoted for your car.

SAFETY PRECAUTIONS

Very high voltages are used in the High Tension side of the ignition circuit, which at the very best can cause an unpleasant shock. While this is unlikely that this would be harmful to a healthy person, any reaction to the shock could well be.

It's therefore wise to take precautions - avoid touching any part of the HT system with your bare hands unless the ignition is switched off. Where it is necessary, such as when checking for a spark, then use insulated pliers, rubber gloves, a dry rag or anything of that nature - do not, for instance, rely on the cable insulation.

These precautions become even more important when working on an electronic ignition system, for a shock from some systems could prove harmful - a shock from any system could have serious consequences for anyone with a heart problem.

It is possible to get a shock from the low tension side of a contact breaker system, where the back emf (electro motive force) of around 300 volts is produced in the primary windings when the magnetic field collapses.

Do not attempt to turn the engine over by hand with the ignition switched on, it's best to remove the key from the ignition in any case.

Some coils and ballast resistors can reach high temperatures in use so avoid touching them both when and for some time after, running the engine.

The electric fans used to cool the radiators in many modern cars, may start up some time after the engine has been switched off, so take care if working in that area. This obviously also applies when the engine is running.

Remember,sparks can start fires and a spark can occur when connecting or disconnecting any lead, or if any live terminal is shorted to earth. Unless it is required for testing purposes always disconnect the battery before working on any electrical component. Incidentally always disconnect the battery negative lead before the positive and connect it up after.

Beware of an items of loose clothing - a tie or loose sleeves when leaning over an engine that is runnning.

CONVENTIONAL IGNITION SYSTEMS

In any petrol burning internal combustion engine the actual combustion process is triggered by a spark jumping across a gap between two electrodes of a sparking plug. The sole purpose of the ignition system is to produce this spark, in the right place, at exactly the right time and of sufficient strength to ignite the fuel/air mixture.

Although the modern trend is to use electronics to switch the electrical currents in the system, there are still a large number of cars around with what are often termed 'conventional' systems where this is done mechanically.

The layout of a typical system of this type as used on four cylinder engine is shown opposite, together with a wiring diagram and a brief explanation of the circuitry involved.

Just as it isn't necessary to know how a car works in order to make a good job of driving it, you don't need to understand how an ignition system functions in order to set it up to the manufacturers specification.

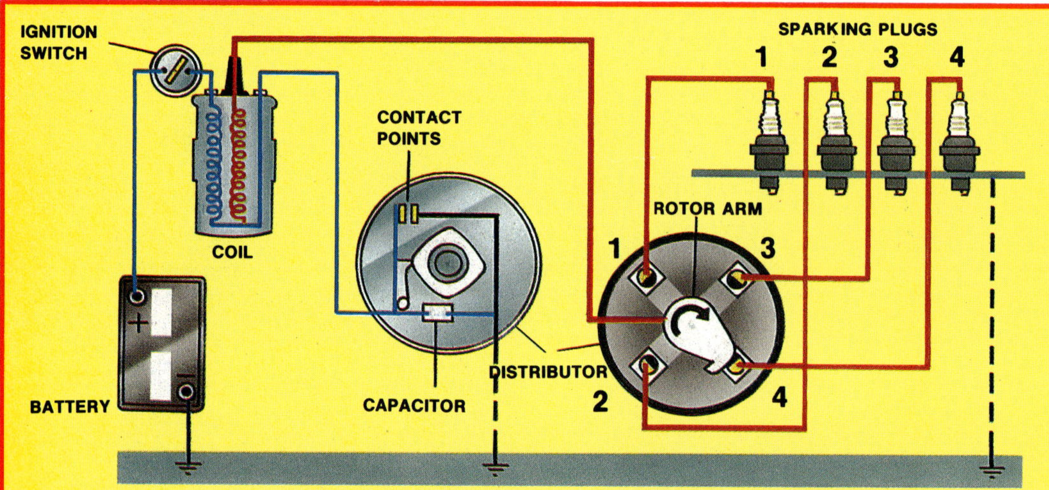

THE PRINCIPLE OF THE IGNITION SYSTEM

What is often termed the conventional ignition system is made up of two separate, but linked circuits - the primary or low-tension circuit (shown in blue) and the secondary or high-tension circuit (shown in red).

Both these circuits are linked at the coil, where the primary winding is made up of about 300 turns or coils of wire and which encloses the secondary winding of around 25,000 turns of much finer wire.

A cam (four lobed in a four cylinder engine) on the distributor shaft, which is driven from the camshaft at half engine speed opens the points four times in every revolution or every two engine revolutions. On each occasion once the lobe of the cam has passed, the points are closed under spring pressure.

When the contact points are closed and the ignition switched on an electrical current will flow from the battery through the primary circuit, resulting in a magnetic field being built up around the primary winding of the coil and which embraces the secondary winding.

Once the rotating cam opens the points it breaks the circuit and the magnetic field collapses. In doing so its lines of force are cut by the thousands of turns of wire in the secondary winding, inducing a high-voltage (15,000 volts or so) electrical pulse into the secondary circuit. At the same time the distributor rotor (mounted on the same shaft as the cam) passes one of the segments in the distributor cap. As it does the high tension (HT) current jumps the gap between the rotor and segment and flows through the HT lead to the sparking plug concerned.

Further rotation of the cam allows the points to close and the action is repeated, but in this case the rotor will be passing the next segment in the cap.

Note: ***the HT lead sequence in the cap corresponds with the firing order, in this case 1-3-4-2.***

In both cases though, it does help and is virtually essential if some part doesn't do what it should. Probably the best way to understand how the ignition system works is to take each component part in turn and see what it does.

BATTERY

This can be considered as an electrical reservoir, being drained off when consumer units such as the lights are switched on and re-filled or rather re-charged by the engine driven generator.

IGNITION SWITCH

For our purposes we can regard this as a simple ON/OFF switch, albeit key operated. In practice the switch is often used for other purposes as well and is generally combined with the starter switch.

COIL

If you have ever watered your garden or washed your car, you will appreciate that, to get a good strong jet, first and foremost, you need high water pressure. In much the same way, to get a good fat spark across the gap in the sparking plug, under all operating conditions can call for high voltages (pressures), in the region of 15 - 20,000 volts.

The role of the coil, therefore is to boost the nominal battery voltage (12volts) to that required at the sparking plug.

In order to understand how the coil does this, it is necessary to have some idea of the relationship between electricity and magnetism. Basically, what this amounts to is that when an electrical current is passed through a conductor (cable or wire) a magnetic field is set up around it.

Many of us, in our younger days at school, may have carried out experiments, where iron filings on a piece of paper, clearly showed the lines of force making up a magnetic field, whenever a magnet or live conductor was brought near.

Not only does a current passing through a conductor, produce a magnetic field but, conversely, if a conductor (wire) is made to pass through a magnetic field, cutting its lines of force, then a voltage is induced into the wire. If the lines of force were cut by a number of turns of wire, then the total

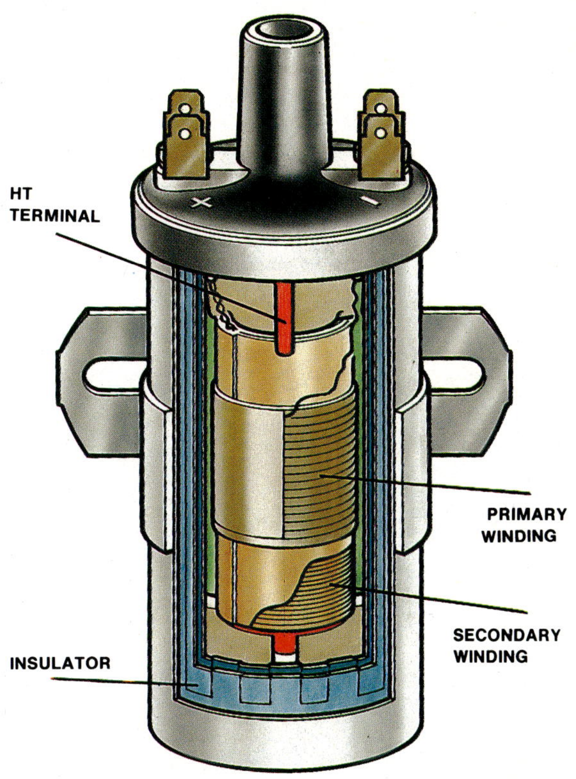

voltage induced would, among other things be related to the number of turns.

Furthermore, if a coil of wire is wound around a soft iron core and a current is passed through the wire, the soft iron core is magnetised, increasing the strength of the magnetic field.

Back then to our automotive coil which consists of two insulated windings and a laminated iron core, all encased in a metal can. Of the two windings the outer (primary) is made up of relatively few (around 300), thicker coils of wire, whereas the inner (secondary) winding consists of around 25,000 turns of much thinner wire.

In operation, battery voltage is applied to one side of the primary winding and the other side is earthed (through the contact breaker). This results in a current flowing through the primary, building up a magnetic field around the iron core. This magnetic field encompasses the secondary winding.

If the current flow is interrupted the magnetic field will collapse and, in so doing, is cut by the thousands of turns of wire in the secondary winding. This induces a very high voltage into the secondary circuit, which eventually ends up at the plug. This is why this system is called inductive.

Each build up and collapse sequence produces one spark yet the engine of a typical car travelling at around 70mph (113kph), would be turning over at some 3,500 revolutions per minute and needing 7,000 sparks per minute.

This means that there must be some means of switching the primary current at this rate and even faster. This device is the contact breaker.

CONTACT BREAKER

Located within the distributor, the contact breaker (sometimes referred to as the 'points') is nothing more than a mechanically operated switch.

Basically the assembly consists of two contacts, one of which is securely attached to the base plate of the distributor - the fixed contact. The other (moving) contact is mounted on a spring strip, together with a fibre compound rubbing block.

This block rides on a cam incorporated onto the distributor centre shaft. In most cases the number of lobes on the cam will equal the number of cylinders. As the cam turns, each lobe will force the contacts

REPLACING CONTACT POINTS

Although this shows a typical contact point arrangement, there can be considerable differences between models. Some Bosch units for example have a bearing plate for the distributor shaft above the points, others may have the centrifugal weights above.

ROTOR ARM

1. The general layout once the distributor cap is removed.

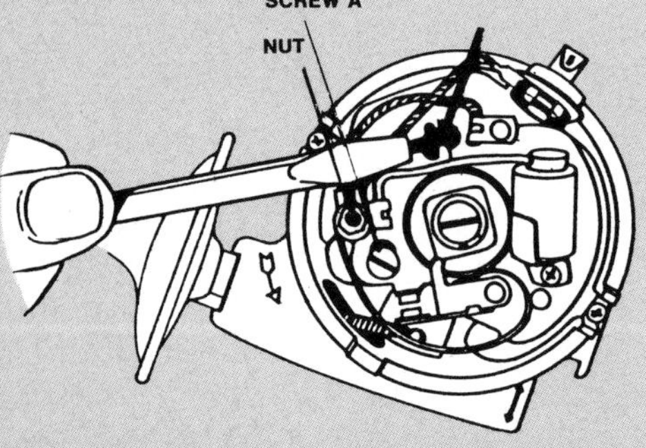

2. The rotor arm just pulls off. To adjust simply slacken screw A slightly and then use a screwdriver in the anchor-plate cut-out (arrowed).

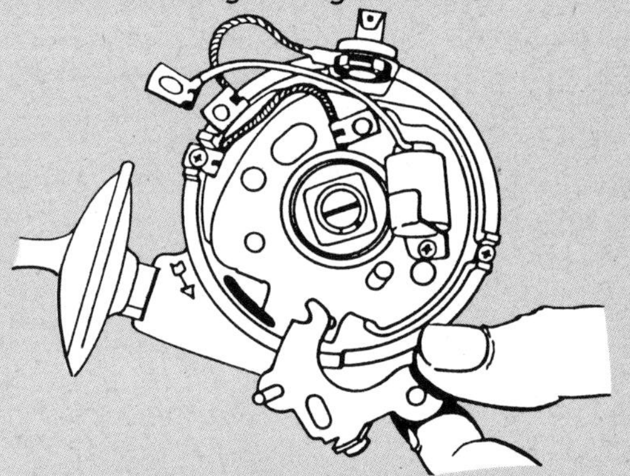

3. To remove the points undo screw A, then undo the small nut on the cable post and remove the cables. Note the location of any insulators. Remove the contact point assembly.

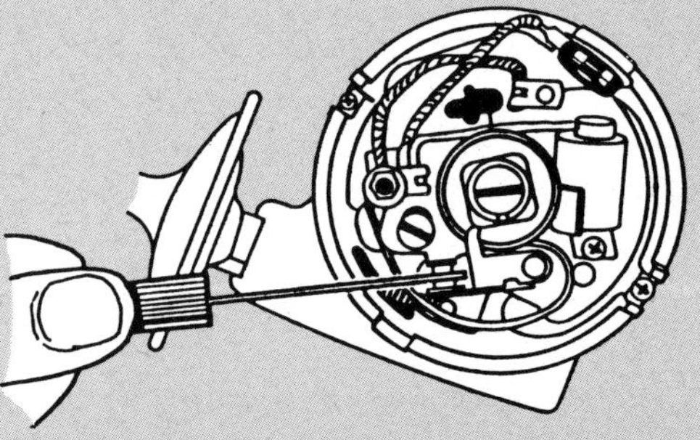

4. Clean off the faces of the new points, put a smear of HMP (High Melting Point) grease on the cam and replace the points. Finally set the gap using a feeler gauge as shown - then check with a dwell meter.

SETTING THE CONTACT BREAKER GAP

The contact breaker gap can be checked with feeler gauges, but even in the hands of an expert this can be very much of a hit and miss affair - trying to measure a gap of around 0.38mm (0.015in) between the two contact faces, when one of them is spring loaded. Even assuming this could be done correctly it doesn't take into account any variations in the cam profiles or of any play in the bearings of the distributor shaft. Ideally the clearance should be checked at each of the cam lobes and any variations averaged out, but this also increases the chances of error.

Another factor to be considered is that unless the contact points are new, some electrical erosion will have taken place, leaving a pit in the face of one contact and a pile on the face of the other and making accurate setting impossible.

The way to overcome this is to check the dwell angle with the engine turning over. This accurately measures the period when the contacts are closed under running conditions, averaging out any variations, for this you need a dwell meter. Having said that owners of many older machines have no choice for only the contact points setting will be given.

apart, after which they will close again under spring pressure.

The distributor shaft is driven from the engine camshaft at half engine speed, this means that, in a four cylinder engine for example, the points would open and close four times every two revolutions of the engine, in other words corresponding with the compression/power strokes of the engine.

Electrically the fixed contact is earthed and the moving contact connected to the CB terminal of the coil. When the points are closed a current will flow through the coil primary, only to be interrupted when the points open.

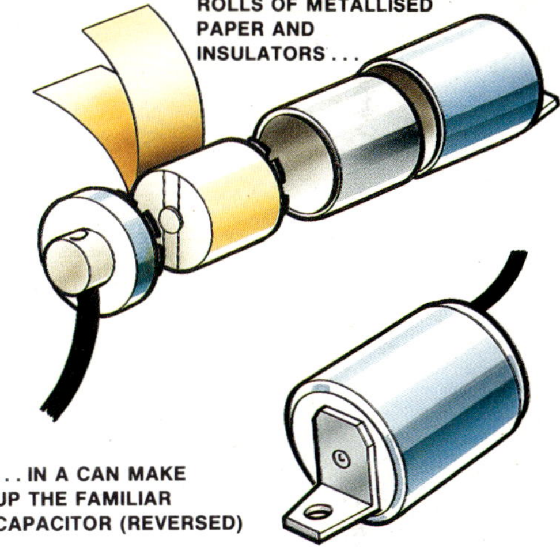

CAPACITOR

When the contacts points open and the magnetic field collapses, its lines of force are not only cut by the secondary winding in the coil, they are also cut by the primary winding, this induces a voltage into the primary circuit. In comparison with the secondary winding the voltage is quite small - somewhere in the region of 350volts - still sufficient though, to try arcing across the just opening contact points.

In a very short time this arcing would burn the contact points to such an extent that they wouldn't pass a current when closed.

The capacitor or condenser as it is sometimes called, is wired electrically across the contact breaker and acts as a sort of sponge, soaking up that surge of current. It is usually located in the distributor.

Having 'soaked' up this current surge, the capacitor, once again acting as a sponge, re-asserts itself and discharges. As the contact points will be by now fully open, the discharged current can only flow back through the primary winding.

This has the effect of speeding up the collapse of the magnetic field and so increasing the voltage induced into the secondary circuit.

The capacitor then reduces arcing at the points and increases coil output.

DISTRIBUTOR

As its name implies the principle role of the distributor is to distribute the HT current from the coil to the appropriate sparking plug. It also houses the contact breaker points and usually the capacitor and, in addition the automatic advance mechanism.

DISTRIBUTOR MAINTENANCE

An often neglected service requirement, which will probably become even more so with electronic ignition systems is lubrication of the distributor. On contact breaker systems this would involve smearing a light film of grease on the cam and oiling the advance weights - this will often also be necessary on those electronic systems with this type of advance mechanism.

Do not over lubricate as any excess oil or grease could be thrown off and contaminate the points or electronic pick-up. While it is off check the inside of the distributor cap for cleanliness and if necessary wipe it out with a clean soft cloth.

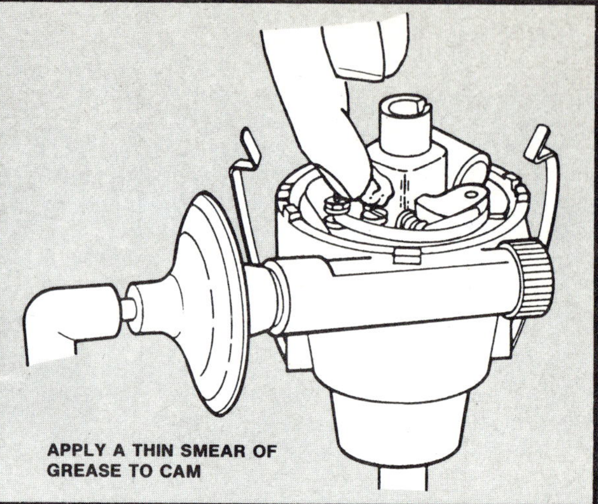

APPLY A THIN SMEAR OF GREASE TO CAM

The high tension section of the distributor consists basically of the cap, which is secured to the distributor body by either clips or screws and the rotor arm, which is fitted over (and turns with) the distributor shaft. HT current from the coil is fed through the centre terminal in the cap, through a brush type connector to the rotor arm electrode.

Evenly spaced around the inside of the cap are a number of electrodes or segments, one for each cylinder. These segments are linked to terminals on the outside of the cap, which in turn are connected to the sparking plugs, in firing order sequence. Both the rotor arm and distributor cap are formed out of a moulded plastic material with very good insulating properties.

The layout is such, that when the rotor arm electrode passes each segment, the distributor shaft cam opens the contact breaker. The resulting surge of high tension current flows from the coil to the rotor, jumps the small gap to the segment from where it moves on to the sparking plug.

AUTOMATIC ADVANCE MECHANISM

If you set fire to a certain amount of petrol it will take a specific time to burn. Much the same could be said for the fuel/air mixture in the cylinder.

In order for the engine to run efficiently it is important that maximum pressure is developed on the piston just as it starts moving down on the powerstroke. For this to happen and to allow for the burning time, the mixture has to be ignited, by the spark, some time before the piston reaches the top of the cylinder on compression stroke (BTDC).

The problem is that, although the burn time of the mixture may be constant, as the engine speed increases, the time (measured in milli-seconds) allowed for it decreases.

This means that as the engine goes faster it is necessary to start burning the mixture earlier, in other words the spark (ignition) has to be advanced.

CENTRIFUGAL (MECHANICAL ADVANCE)

To achieve this automatic advance of the spark, the distributor shaft is made up of two parts, the lower section is driven from the engine camshaft, while the upper part incorporates the cam and drives the rotor arm. the two are joined by an arrangement of pivoted weights and springs.

In use as the engine speed increases, the weights move out against spring pressure, moving the upper part of the shaft and therefore the cam in relation to the lower shaft. This (forward) movement of the cam means that the points open and close earlier - in effect, advancing the spark.

The relationship between the size of the weights and the strength of the springs determines how far the weights move out and therefore the degree of advance at any given engine speed. This 'advance curve'

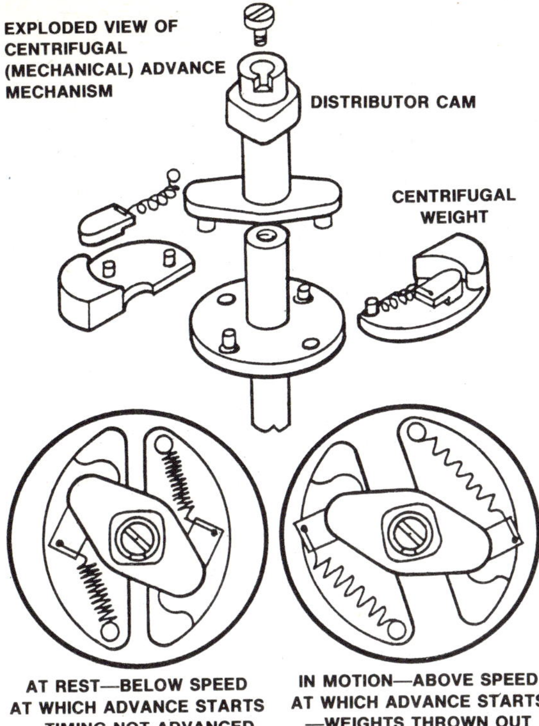

varies from engine to engine but in a typical example, centrifugal advance would start at around 400rpm and finish around 2,400rpm. Note that this is distributor shaft speed, the engine would be running twice as fast.

VACUUM ADVANCE

Although it was stated that the burn time of the fuel/air mixture was more or less constant, this only applies if the volume is also constant. However in an engine the amount of air (and fuel) going into the cylinders changes, according to the position of the throttle.

For instance, when the throttle is only partly open, less air enters the cylinders than when it is fully open. This means that the pressure rise on compression stroke is also reduced.

The fuel/air mixture burns more slowly at lower pressures so, in these conditions, the ignition needs to be advanced even further.

This is achieved by the use of a vacuum control unit, usually mounted on the side of the distributor and operated by the depression felt at the venturi in the carburettor. This arrangement is such that, as the diaphragm within the unit flexes, it physically moves the base-plate on which the contact breaker points are mounted.

By moving the contact breaker in relation to the cam the contacts are made to open earlier, so advancing the spark.

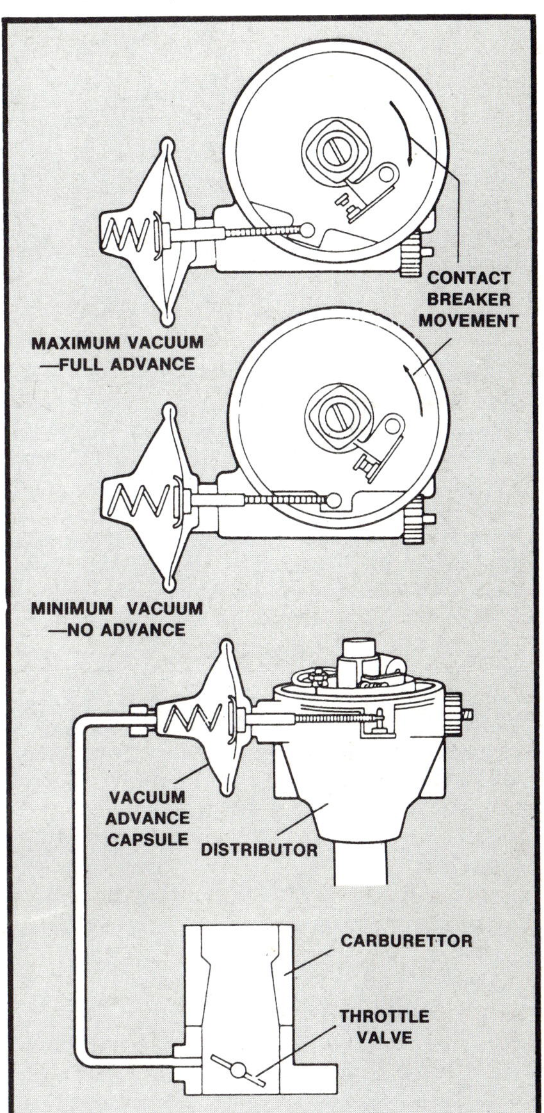

VACUUM RETARD

In some cases to comply with stricter emission control regulations ignition systems may be fitted with a vacuum retard unit, designed to retard the ignition at very high manifold readings, such as on over-run (closed throttle deceleration), when there could be a brief enrichment of the mixture.

BALLAST RESISTOR

When the starter is operated, it makes a big drain on the battery, in effect reducing its terminal voltage by anything up to 3 or 4 volts. In turn this reduces the coil output and weakens the spark.

To overcome this, a coil designed to operate at around 8 volts is used and a (ballast) resistor inserted into the circuit between the ignition switch and coil, so as to bring the nominal 12 volt battery output down to that required by the coil.

This resistor is by-passed when the starter motor is in use so that the consequent reduction in battery voltage, has no effect on the spark.

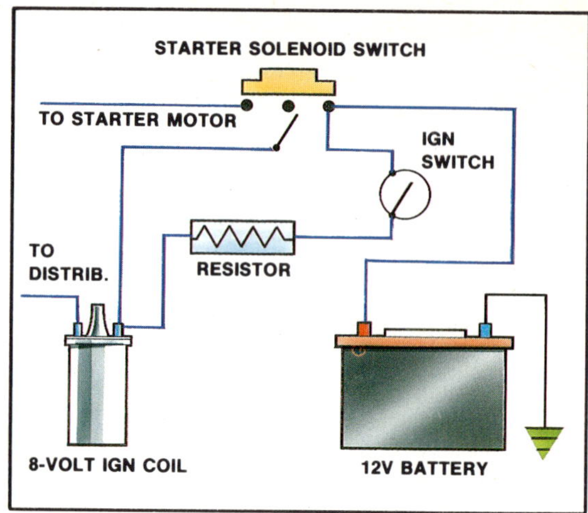

DWELL PERIOD

Basically the dwell period is that part of the distributor cams rotation, when the contact breakers points are closed and can be expressed either as an angle (°) or a percentage (%).

A four lobe cam, for instance, rotating through 360 degrees, will do so in four periods or phases of 90 degrees. In each of these periods the points will be closed for some of the time - this is the dwell period and is critical because this is when the coil is being 'charged' up.

In a typical four cylinder engine (four lobe cam) the points would be closed for around 54 degrees in each 90 degree period. The dwell angle is then 54

In percentage terms this equals

$$\frac{54}{90} \times 100 = 60\%$$

To convert back to degrees

$$\frac{60}{100} \times 90 = 54°$$

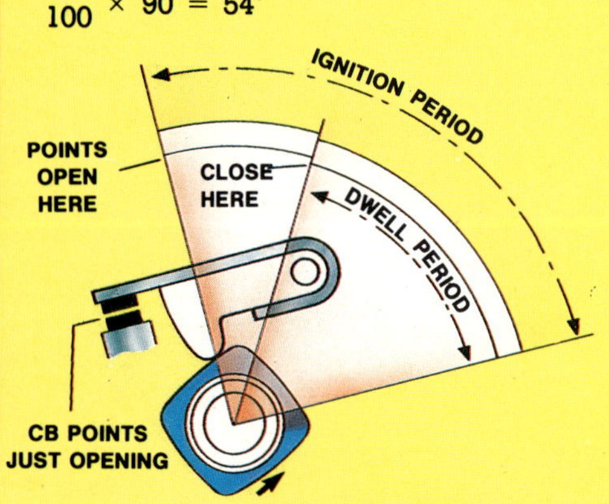

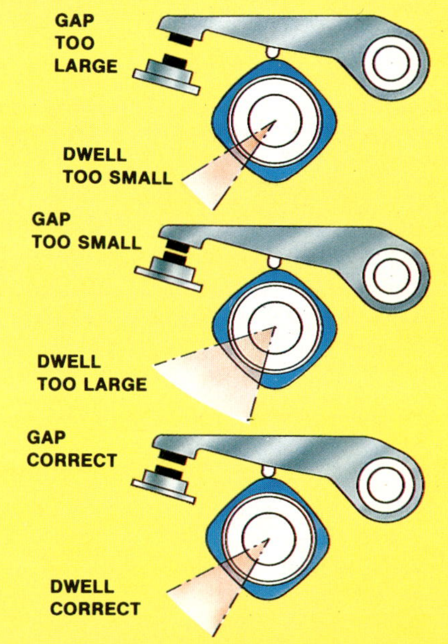

The contact breaker gap and the dwell angle are related in that the wider the gap, the longer the points remain open and the shorter the dwell angle. The opposite applies if the gap is smaller than specified.

An incorrect gap or dwell angle will also have a direct effect on the ignition timing. For instance in a typical four cylinder engine with a dwell angle of 54 an actual setting of 60 would mean that the contacts opened 3 degrees late, but as the distributor shaft is driven at half engine speed, this is equivalent to 6 degrees at the flywheel. The ignition would therefore be 6 degrees retarded.

ELECTRONIC IGNITION

Although an electronic ignition system can be designed to provide a stronger spark than a conventional system, the main reason for its greater efficiency is that it stays in tune much longer.

The conventional system is at its best shortly after a new set of contact points have been correctly fitted and gapped. From then on as erosion and wear takes place efficiency tails off. Ignition timing is also effected, for as the heel wears, the points gap decreases and the ignition becomes retarded.

Although it may initially go unnoticed, this deterioration of the contact points can cause a high speed misfire, which results in unburnt petrol being pushed out into the exhaust pipe, producing high emission levels. Retarded ignition will make this even worse.

In order to ensure that their engines comply with emission control regulations, most manufacturers now fit electronic ignition systems which eliminate the contact breakers. This along with other factors has reduced the maintenance requirements allowing the same manufacturers to extend the service intervals.

This, however is only a transitionary stage for more and more manufacturers are now fitting ignition systems which incorporate electronic control of the ignition advance as part of an engine management system. In some cases the distributor has also been replaced by a multiple coil system, in which the ignition process is triggered off by sensors on the flywheel.

In general, electronic ignition systems and especially the Capacitor Discharge versions produce a higher secondary voltage than conventional systems. This could be extremely dangerous, even fatal if any exposed terminal or live part is touched. It is always advisable then to ensure that the ignition is switched off before working on, or checking any part of the system. In addition with some systems the electronics can be damaged if a disconnection is made with the engine running.

TRANSISTOR ASSISTED CONTACT (TAC) SYSTEMS

Still available as an after-market fitting, these TAC systems retain the contact breaker points, but use them to trigger the electronic process, whereby a transistor is used to switch the coil primary current.

This completely eliminates electrical

DIGITAL IGNITION SYSTEMS

Digital ignition systems (sometimes referred to as 'map' ignition) incorporate a microcomputer which calculates the point of ignition and the necessary timing advance or retard. It does this using three dimensional maps stored in its memory and information fed in from various sensors.

This information may include details of engine temperature, outside (ambient) temperature, throttle position, engine speed and the onset of detonation (knock).

Digital systems provide a much more accurate method of timing the spark and are not affected by wear of mechanical components. Efficiency is therefore improved, exhaust emissions are reduced and less fuel is used.

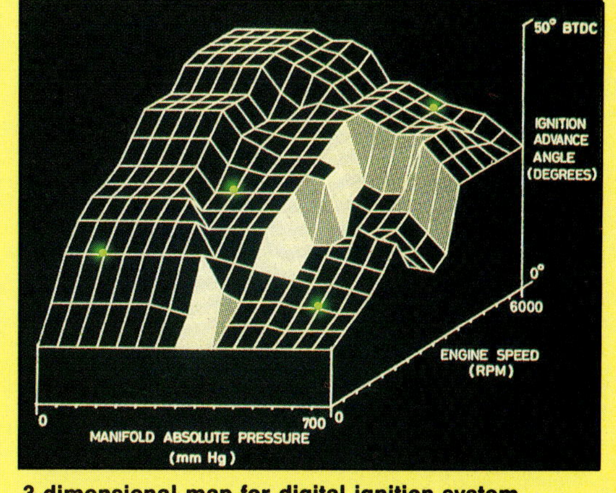

3-dimensional map for digital ignition system

erosion of the points for instead of switching the three to five amps of the coil primary they now only switch less than 1/3rd of an amp, sufficient to keep the points clean but not enough to cause any arcing. In any TAC system the contact points should last at least 30,000 miles.

In an ordinary contact breaker system, the dwell period (when the points are closed) gets shorter as the engine speed increases, whereas with the better TAC systems the only function of the contact points is to signal to the electronic control unit that a spark is required. Just as in a conventional system this is when the points open. Unlike conventional systems however, where the coil is switched on when the points close, in these TAC systems it is done electronically.

This can help give the coil a longer 'charge time' particularly at higher engine speeds. In this sense then these TAC systems can give an improved spark, when compared with a conventional system in good order.

Advantages of these systems over the contact-less versions include:

1. Much cheaper to buy.
2. Relatively easy to fit.
3. Easy to change from one vehicle to another.
4. Easy to change back to conventional in the event of electronic failure. Many systems have a change over switch.
5. If the distributor and its drive gears are worn, removing the contact breakers points also removes the pressure and drag they put on the shaft and which helps take up any free play. Removal of the points then can result in the shaft oscillating with consequent timing errors at individual cylinders.

Disadvantages include:

1. Although points erosion may be eliminated, wear - particularly on the fibre heel is not, so the points will still need periodic adjustment. As with conventional systems, the timing will alter as wear takes place.
2. The problems of 'points bounce' in high-speed applications can still arise.

TRANSISTOR ASSISTED CONTACT (TAC) ELECTRONIC IGNITION SYSTEM LAYOUT

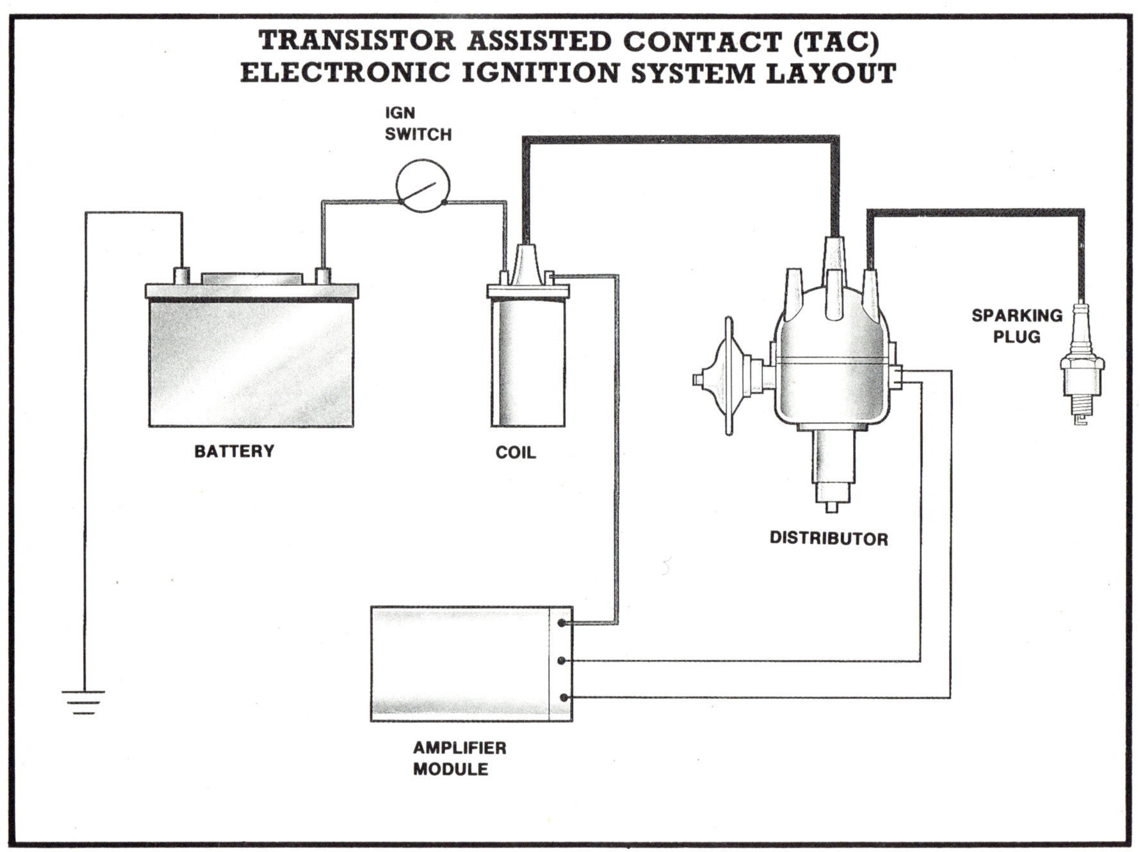

CONTACT-LESS SYSTEMS

As the name implies these systems eliminate the contact points completely. With many the coil primary current is still switched by a transistor but it is the method of triggering the electronic process that is different. There are in fact three different methods in use - optical, magnetic and what is termed the Hall Effect. All three are available in conversion kit form, for modifying existing contact breaker systems, but only the latter two are used in the factory fitted systems.

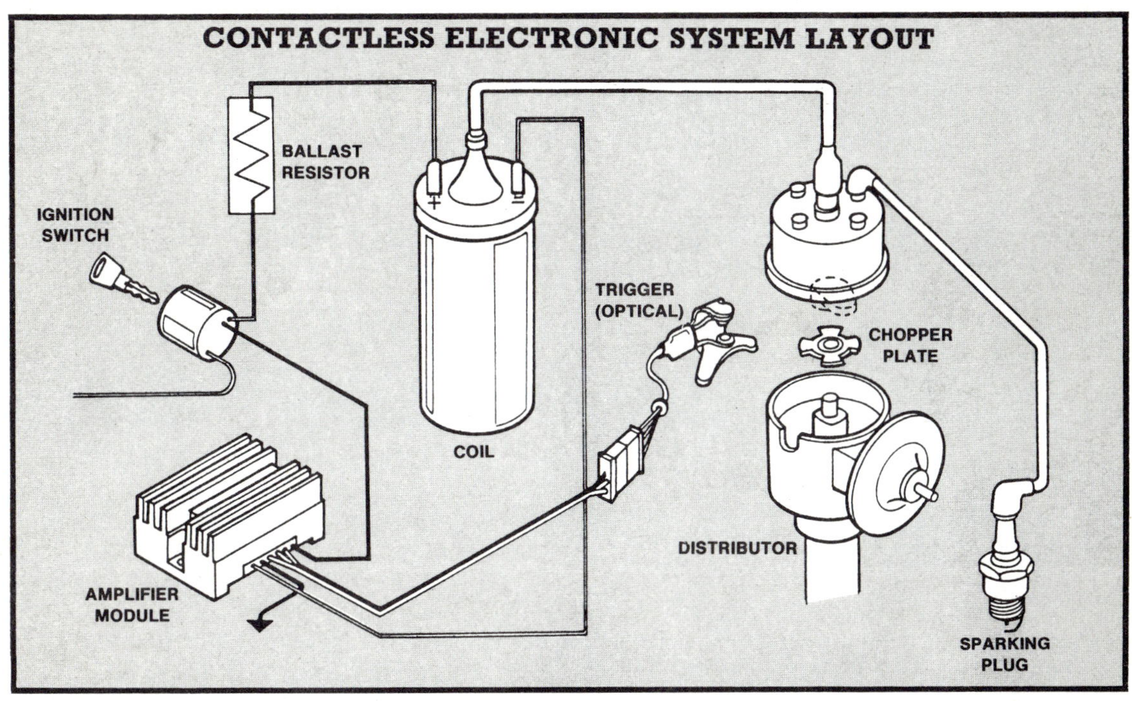

OPTICAL SYSTEMS

In these systems, a horseshoe shaped bracket containing a light emitting diode in the end of one leg and a phototransistor directly opposite in the other is fitted in place of the contact points and capacitor. When the ignition is switched on a beam of light is projected from the diode onto the phototransistor.

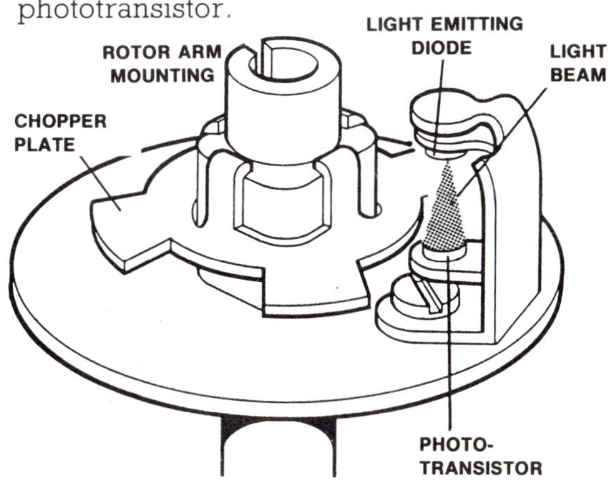

A segmented disc or chopper plate is mounted on the distributor shaft so that as the disc rotates (with the shaft), its blades interrupt the beam of light, producing a signal which is then used within the control module to switch the primary current and produce a spark. As the number of blades on the chopper disc corresponds with the number of cylinders it acts in much the same way as the cam in a conventional system.

MAGNETIC SYSTEMS

Magnetic systems vary but a typical arrangement would consist of a spoked iron rotor mounted on the distributor shaft, a pick-up unit fitted in place of the contact breaker points and capacitor and a control module.

The pick-up unit consists of a permanent magnet and a small inductive winding (coil), making what is in effect a pulse generator.

As the iron rotor turns (with the distributor shaft), each spoke in turn passes close to the pick-up unit. This produces a change in the permanent magnetic field, which induces a small electrical current into the winding and is used as a signal to the control module in much the same way as the optical version. The number of spokes on the rotor corresponds to the number of cylinders.

The air-gap between the rotor spoke and pick-up should never alter, despite being

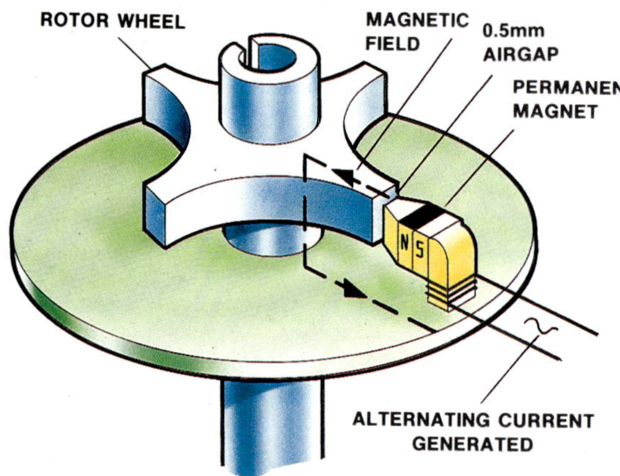

shown as a service check in some manuals. Only use feeler gauges made of a non-magnetic material, when checking the air-gap.

Another form of magnetic triggering uses ferrite rods (small rods of high-permeability magnetic material), set in the rotor and which have a similar effect on a pulse generator as the rotor spokes.

HALL EFFECT

Named after its discoverer, the American E. H. Hall, this works on the principle that if a chip of semi-conductor material (silicon chip) is located in a magnetic field and has a current passed through it, a small voltage is generated in the chip at right angles (physically) to the current.

In practice the chip and a permanent magnet are mounted on the distributor base plate in place of the contact points and capacitor, with a gap between the two of them. A slotted metal vane wheel is attached to the distributor shaft, in such a way that as the shaft rotates, the metal vanes pass between the permanent magnet and chip. Every time this happens the Hall

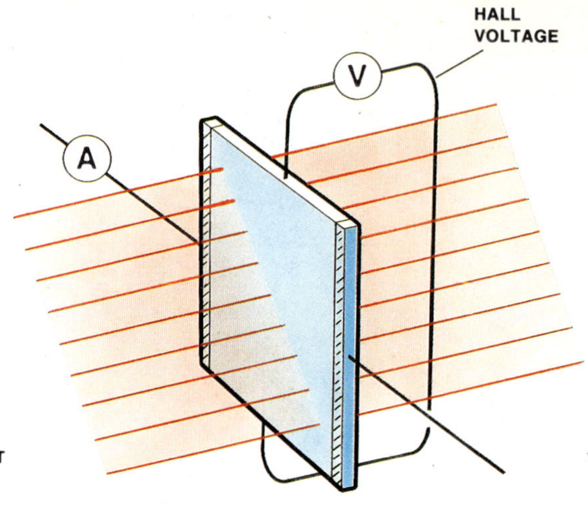

voltage is generated in the chip and used in the control module to switch the primary current. The number of vanes corresponds to the number of cylinders.

Unlike some other systems where the dwell period is determined by the control module electronically switching on the primary current to the coil, the vanes in the Hall Effect units switch the primary both on and off. The dwell is therefore related to the width of the vanes and is constant throughout the engine speed range.

CAPACITOR DISCHARGE (CD) IGNITION

All the electronic ignition systems mentioned so far work on the inductive discharge principle, in that the collapse of the magnetic field induces a voltage into the secondary windings of the coil. this means that the coil has to be charged up over the dwell period and that the energy for the spark is then stored in the coil, until it is released by whatever triggering system is used.

The CD ignition system works on an entirely different principle in that the ignition energy is stored in a capacitor which, when a spark is required discharges through the primary winding of a coil. In this application the coil doesn't store any energy but acts as a transformer, transferring the energy from the primary to the secondary winding.

Basically with CD systems, battery

voltage which would normally power the coil is used through a transistorised converter to charge a capacitor at around 300 to 400 volts. When a spark is required a special type of solid-state switch, called a thyristor discharges this voltage through the coil primary, producing a high energy spark, which is, however, only of very short duration.

The high tension circuit in CD systems is the same as in all the others and any of the triggering methods can be used to switch the thyristor.

result in a failure in an inductive system.

The CD systems main disadvantage is the duration of the spark. In an ideal situation this would be in the region of 0.2 milliseconds compared with a minimum of around double that from an inductive system. In theory this short high energy spark would be ideal, in that once having ignited the mixture any additional spark length would be superfluous. In practice though, the fuel/air mixture may be mixed unevenly and the ignitable part of the mixture may not have reached the plug

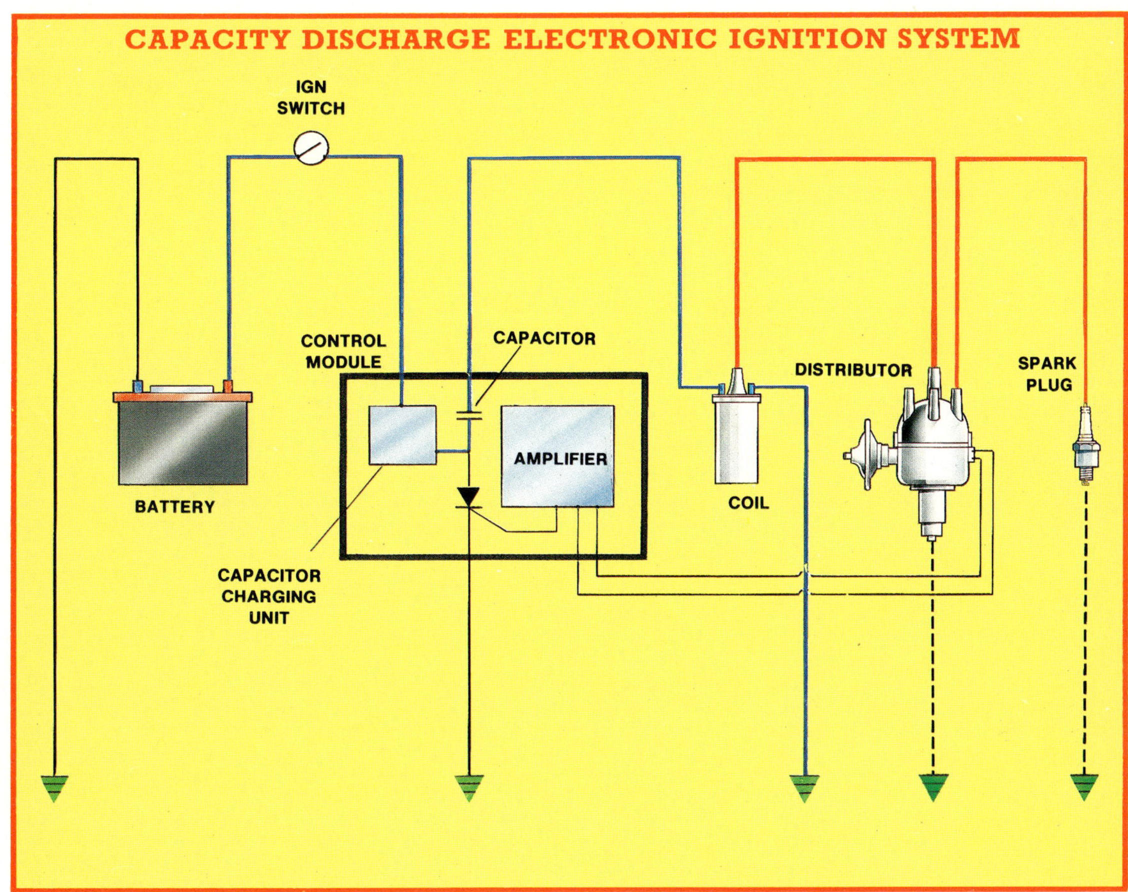

The main advantage of the CD system is that the secondary voltage rise time is around ten times faster than in any of the inductive systems, it is therefore capable of producing many more sparks per minute. For this reason it is mainly used in high speed multi-cylinder engines. A further point in its favour is that the very high secondary voltages tend to blast through any shunts. such as fouled plugs, that would

when the spark occurred. This would be even more probable on those engines designed to run on a weak mixture.

A number of these systems use extra electronic circuitry to extend spark duration or, rather, produce a series of sparks. The problem with most of these, however, is that the follow up sparks can be much weaker than the original. So weak, in fact, that they may fail to ignite the mixture.

COILS FOR ELECTRONIC IGNITION SYSTEMS

Converting a conventional ignition system to electronic may in the long term prove beneficial both in terms of fuel economy and starting performance, but in order to derive maximum benefit, the coil should be replaced with a higher powered equivalent.

All things being equal a coils output is dependant upon the ratio of turns of wire in the secondary winding to that in the primary, the higher the ratio, the greater the output. There are two ways of increasing this ratio - either increasing the number of secondary windings, which increases the size and weight of the coil or cutting the number of turns in the primary, which has the opposite effect.

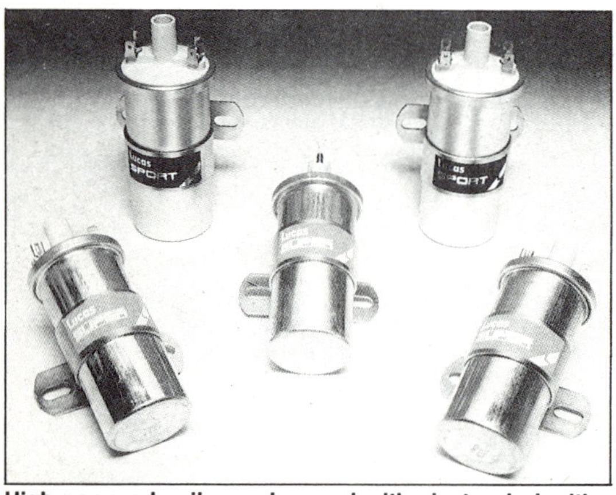
High powered coils can be used with electronic ignition

Most high performance coils fall somewhere in between, but reducing the number of turns in the primary winding will result in a greater current flow, possibly in the region of 7 amps or so, compared with standard coils 3-4 amps. This higher current flow would be far too much for an ordinary contact breaker, but is well within the limitations of an electronic system, whether TAC or contact-less. A fewer number of turns in the primary winding also reduces the back emf (electro motive force) produced in the primary when the magnetic field collapses. this allows a quicker rise time for the coil current and in that way helps increase the efficiency of the coil.

Furthermore if the original (contact breaker type) coil is used, it must be remembered that the voltage drop across the switching transformer is greater than that across a set of points in good condition. This could conceivably result in a standard coil producing a lower output when used with an electronic system than it did with a conventional one.

FAULT FINDING ON ELECTRONIC SYSTEMS

Although the basic operating principle of many electronic ignition systems may be similar in a large number of cases, the layout of the systems vary enormously. It is therefore difficult to give any detailed and specific procedure.

Basically any checks of the High Tension side of the system should follow a similar pattern to that for a conventional system, except of course in most cases the contact points cannot be flicked open when checking for a spark. However with optical systems it should be possible to interrupt the light beam, even if just with a credit card or something similar, much the same can be done in those magnetic and Hall effect systems using a metal vane or magnet. The alternative is to crank the engine over.

Other than with CD systems it should still be possible to check for a 12volt supply at the coil - this again should be intermittent as the engine is turned over. Depending on the layout and the type of connections, the supply to the control module could also be checked.

Although there are a large number of different wiring layouts, with some being part of an engine management system, basically with most the ignition side will consist of the control module and the triggering device. The control module will have a power supply cable, a cable to each side of the coil and two cables to the triggering device. Some may also have an earth cable, while with others the earth may be through one of the units mounting screws.

On most it should be easy to check the power supply and the power at the coil terminals. On some it may also be possible to check the signal power from the triggering device, but you would need a voltmeter capable of picking up readings of around 0.2 volts eg. Gunson's Testune. It may be possible to use the 'points condition' mode of some multi-meters for this purpose.

In the majority of failures with electronic systems, problems with the various connectors are the most common.

Take care when working on electronic ignition systems, especially the CD versions. Some systems can be damaged by disconnecting an HT cable with the engine running, if so it should be stated in the vehicles handbook.

In some magnetic triggered systems, there may be an air-gap measurement given between the pick-up in the distributor and the rotor vane, and which should be checked occasionally. Under normal circumstances this should never alter, but only use non-magnetic (plastic) feeler gauges when carrying out any such check.

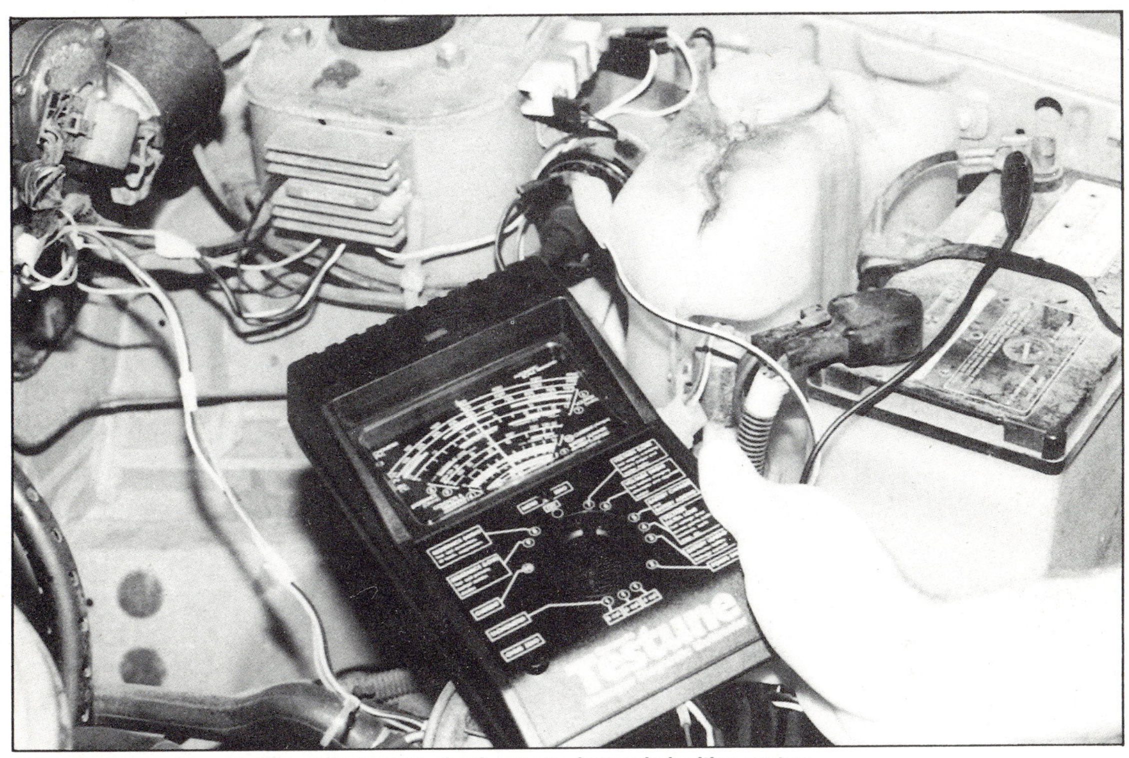

Using the low voltage facility of Testune to check out an electronic ignition system

IGNITION TIMING

If the spark occurs later than it should, the whole combustion process will be late and the piston will be already part way down the cylinder when maximum expansion of the burning gases takes place. This means that the pressure exerted on the piston will be less, with a consequent drop in power output.

In addition the gases will still be burning and producing heat when the exhaust valve opens. Much of this heat will be transferred into the cylinder head around the exhaust port and the engine will overheat. In time it would also burn the exhaust valve.

A spark which occurred earlier than it should can cause even greater problems, for in this instance it could be that maximum expansion takes place before the piston has reached the end of compression stroke. This can result in detonation or as it is more commonly called 'pinking' - a metallic tinkling noise from the engine, particularly noticeable when accelerating and which, in severe cases can cause extensive engine damage.

Correct timing then is essential, but as the contact breaker setting will effect the timing it's essential that this is first checked and if necessary adjusted. Too wide a gap will make the points open late so retarding the spark, whereas too small a gap will have the opposite effect.

With either method of checking the timing (static or dynamic) it is first necessary to know where the timing marks are, what they represent and what the manufacturers recommended settings are.

On the majority of cars the timing marks will be on the crankshaft pulley with a corresponding mark or pointer mounted on the engine block, adjacent to the pulley. Alternatively it could be just a V notch in the pulley and a series of marks on a plate fixed to the block. A selection of different layouts is shown below.

On some cars however, the marks are on the flywheel and can be seen through a (covered) window in the flywheel housing. With many of these it can be advantageous to use a mirror in order to gain a better view of the marks.

If the timing is found to be incorrect regardless of which method is used for checking, it can be corrected by releasing the distributor clamping screw a little and turning the distributor body. Turning it in the direction of rotation of the rotor arm will retard the spark whereas turning it against the rotor will advance it. Some older distributors may have a knurled wheel type of fine adjustment provided.

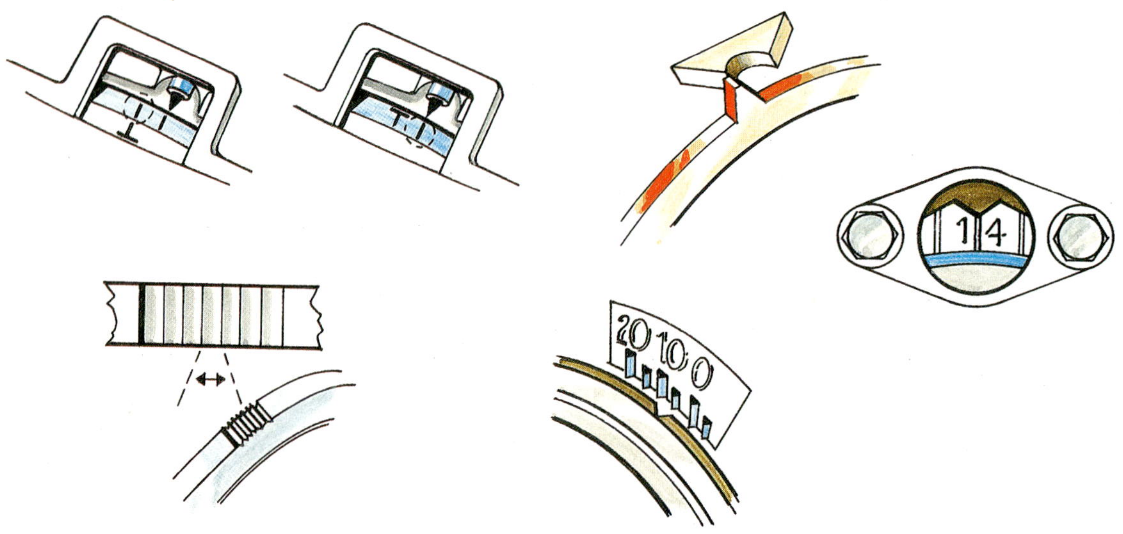

STATIC TIMING

If a test lamp or voltmeter is available it should be connected between the distributor input terminal (coil to distributor lead) and earth. Then with the ignition switched on and the distributor cap removed, turn the engine in its normal direction of rotation until just before the timing marks align and the rotor arm is approaching number one segment (if the cap were fitted)

The test lamp will light up exactly when the points open, which is when the spark occurs. So if the engine is turned over a little further very slowly, the lamp should come on as the marks align.

If the timing is incorrect the easiest way to remedy the situation is to align the timing marks, then slacken the distributor clamp screw and rotate the distributor body a little in the direction of rotation of the rotor, then bring it back slowly until the test lamp comes on. The setting will now be correct.

An alternative method on most four cylinder engines is to remove the plugs and then re-connect plugs one and four to their respective leads. Now with the ignition switched on, turn the engine until either of the plugs fire when if the timing is correct the marks should align. Beware of sparks from the disconnected plug leads, if using this method. A further alternative is similar but instead of removing all the plugs, remove just number one and fit a Colortune glass topped sparking plug, then look through it to see the spark - full instructions are supplied with the kit.

DYNAMIC TIMING

When checking the ignition timing dynamically all that is necessary is to connect up the strobe according to the directions supplied. If a neon version this will simply mean connecting it between the plug (usually) at number one cylinder and its plug lead. Xenon tube types vary but many will need an outside power source - either a mains supply or a couple of connections to the car battery.

The engine should then be started, adjusted to run at the specified speed and the strobe pointed at the timing marks. with its trigger (xenon versions) pressed.

If the timing is correct, the stroboscopic effect should make the appropriate marks align.

In the majority of cases this check should be carried out with the vacuum advance pipe, disconnected at the distributor end and plugged.

It can make things easier, especially with the less bright neon strobes if the appropriate marks are first cleaned up and made to stand out better by the application of a dab of white paint or chalk.

Although this checks the basic timing, normally at or around idling speed it is no guarantee that the timing is correct as the engine goes faster and the automatic advance mechanisms come into use. Some manufacturers now specify a timing check at around 3,000 engine rpm as well as at idling. These tests can only be made with equipment such as Gunson's Supastrobe (See page 12).

This picture of timing marks was taken with the engine running and using Supastrobe as a flash

SPARKING PLUGS
and H.T. CABLES

Although a sparking plug may look a fairly simple component and only has to provide a gap between two electrodes for a spark to jump across, in reality it has to operate in a pretty hostile environment and it's a credit to the manufacturers that they last as long as they do.

In the first place, it has to act as an electrical insulator for voltages up to in some cases 30,000 volts. In addition it has to withstand pressures from below atmospheric to over 1,000psi (69 Bar) and temperatures of below zero to around 2,000 degrees, yet once the engine is running, the firing end of the plug needs to stay in a fairly narrow temperature band of from 350 to 850°C. Furthermore it has to withstand constant vibration when the engine is running and chemical attack from the combustion process and additives in the petrol.

A typical plug consists of a metal electrode passing down the centre of the insulator, which is generally made from aluminium oxide. Surrounding the bottom of the insulator is the metal shell or body, often zinc plated to protect against rust and chemical attack. Welded to this casing, and therefore earthed to the engine, is the side or earth electrode, separated from the lower end of the central electrode by a small gap.

There are two seals required in each sparking plug, one between the central electrode and insulator and the other between the insulator and shell. Some manufacturers use their own specially formulated 'cement' for this while others use various forms of caulking rings.

In some plugs a resistive central electrode may be used to reduce ignition system interference with radio and television.

HEAT RANGE

For optimum performance, the plug firing end temperatures shouldn't rise above 850° even under continuous high speed conditions, neither should they fall

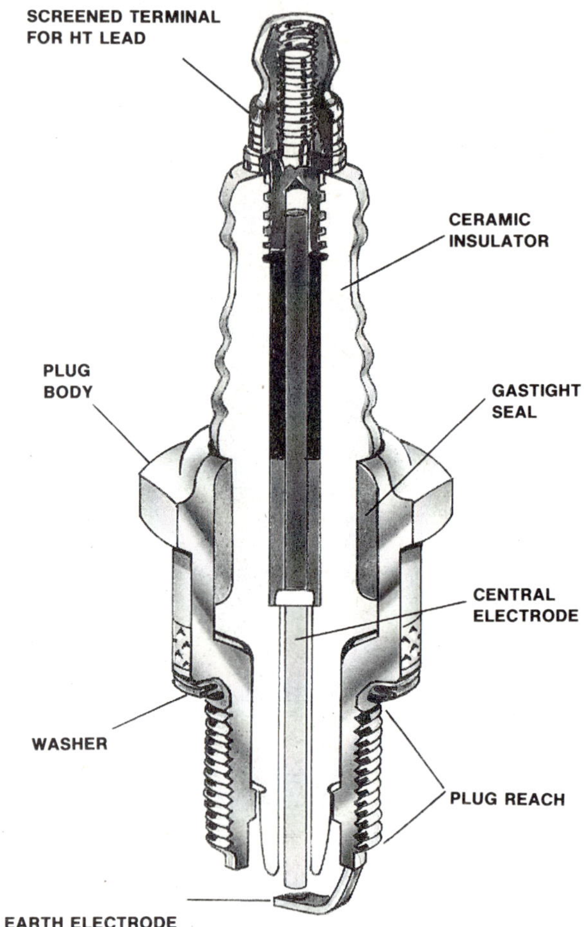

below 350° or so under 30mph cruise conditions.

At temperatures below 350° the plug will not be hot enough to burn combustion deposits off its central insulator (something like a self-cleaning oven) while at much above 850° there is a danger of electrode burning and of the tip of the plug glowing incandescent red. This could fire the mixture before the spark occurs a phenomenon known as pre-ignition or pre-detonation (see page 43), and which can result in severe engine damage.

Combustion temperatures may reach around 2,000°C but engine designs and therefore combustion chamber shapes vary, along with the actual location of the plug in the head. This means that the heat absorbed by the sparking plug may vary from engine to engine . For this reason a plug that may be suitable for one engine may be no good for another, even assuming it would fit. All plug manufacturers therefore produce a range of plugs graded according to their heat dissipating qualities.

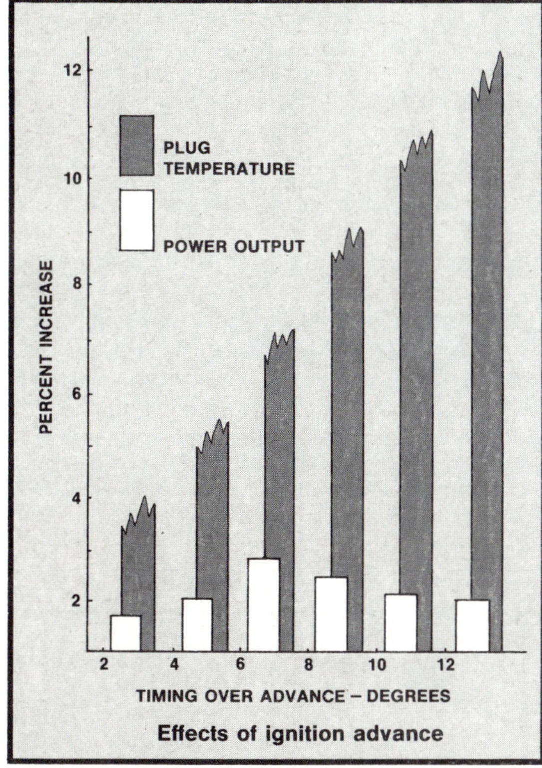

Effects of ignition advance

Although the sparking plug is cooled to some extent by the incoming fuel/air mixture, once during each cycle, around 90% of the heat it absorbs is lost through conductivity into the cylinder head. The heat path is then through the insulator to the body or shell, then on into the engine. The shape of the insulator, therefore determines how quickly heat is lost.

In general a high compression engine runs hot and needs a plug with good heat dissipating qualities, whereas a low compression engine runs cool and needs a 'hot' plug. The rule is therefore:

A hot engine needs a cold plug
A cold engine needs a hot plug

Some people refer to a cold plug as 'hard' and a hot plug as 'soft'.

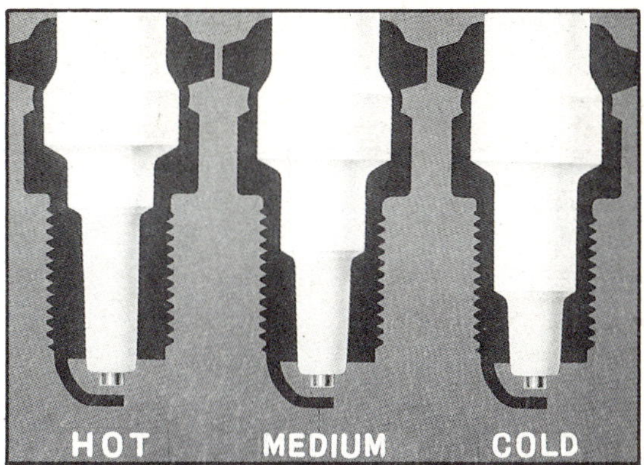

One number in the sparking plug specification denotes its heat qualities, for example with Champion plugs the heat range extends from 1 to 25, the most popular being the 9, as in the N9YC plug used in a large number of cars

PLUG SIZES

Sparking plugs have always been available with different thread sizes by far the most common are those with a 14mm diameter thread, however these can have either a 21mm or 16mm, across the flats (AF) hexagon, each requiring a different sized spanner. However some cars may use plugs with an 18mm thread diameter, whereas plugs with 10 or 12 diameter threads may be found on motorcycles.

In addition the length of the threaded portion can be either 3/8in (9. 5mm), 1/2in (12. 7mm) or 3/4in (19mm). Should a shorter than specified plug be used, the plug tip is shielded and may not ignite the fuel/air mixture, even if it does there is a chance that the screwthreads left exposed may be damaged by heat. Using a longer than specified plug could result in it being struck by the ascending piston, causing considerable engine damage.

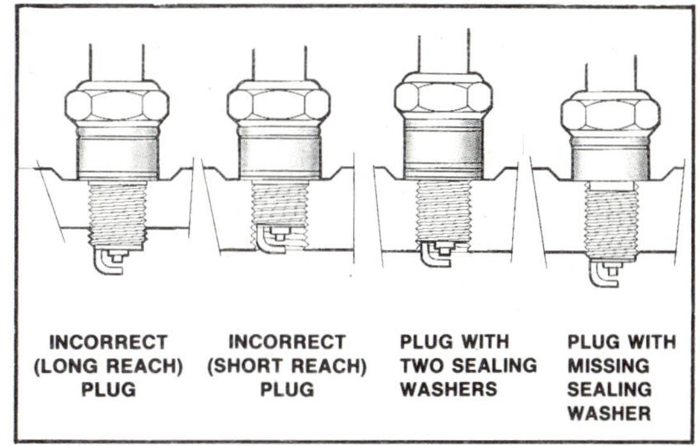

SPARKING PLUG MAINTENANCE

Most manufacturers recommend that the sparking plugs are replaced at around 12,000 mile intervals, but that is with an engine in good condition and even then it's advisable to check, and if necessary clean and reset them at the half-way mark. With older engines which may be burning oil or may be running with a rich fuel/air mixture, it's advisable to check a little more frequently. In fact the plugs can be a good indication of engine condition as shown in the first book of the series (Simple Engine Tuning).

As a general rule of thumb, it can be reckoned that electrical erosion of the central terminal is about 0.001in every 1,000 miles.

Before removing the plugs from the engine, number the leads to ensure that each goes back onto the correct plug. Use either a length of sticky tape or the little tags that some bread producers use on their sliced loaf packets.

Avoid disconnecting sparking plug leads by pulling on the lead itself. Most now have some form of insulator cap which fits over the plug, where possible pull on this cap.

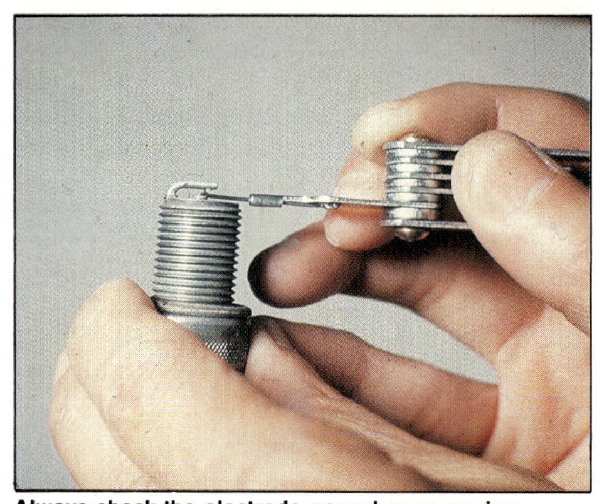

Always check the electrode gap when removing the sparking plugs. Bend the outer electrode to obtain the correct setting

Again where possible, use a soft paintbrush to displace any debris which may have fallen into the recess around the plug and which could otherwise drop into the engine when the plug is removed.

When removing the plugs with either a socket or box spanner, take care to keep the tool straight and in line with the plug.

HT CABLES

These are of two types - copper cored, which should be fitted with suppressor caps at the plug end to eliminate any interference on radios and televisions, and the now almost universal graphite cored versions which are resistive and self-suppressing.

After some time these graphite cored cables can break down and become practically open-circuit, resulting in a misfire. They can be checked with an ohm-meter (multi-meter), in some cases the resistance per linear foot is marked on the cable insulation, but in any case expect the average length cable to have a resistance of up to 20,000ohms.

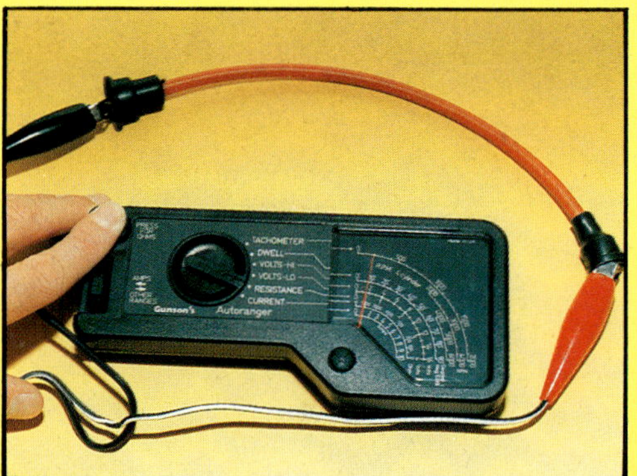

Defective HT cables can affect performance. Check them with a multi-meter in the ohms mode

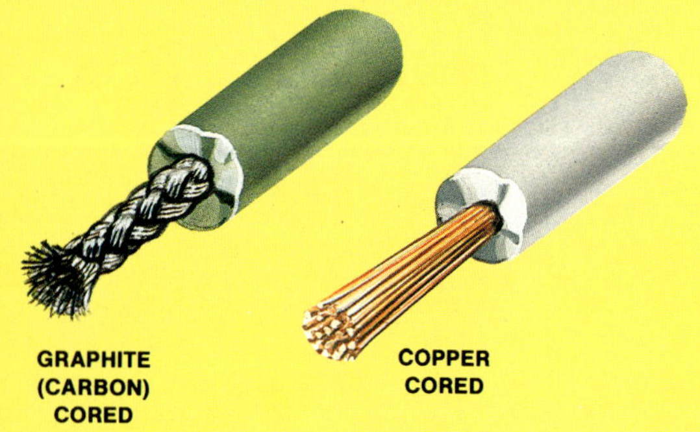

Avoid bending graphite cored cables through sharp angles and do not pull on the cable itself when detaching it from the plug. If the cables become dirty or oil-fouled clean them off with a rag and methylated spirits.

Failure to do so could result in the plug being damaged. The better plug spanners incorporate a rubber insert to prevent this happening - this type of spanner is also preferable when fitting plugs to those engines where access is a problem.

The best way to clean a plug is by sand-blasting, although the electrodes can be scraped off with a brass wire brush or even with a knife or flat file. Take care not to scratch the insulator, since this may lead to 'tracking' when the high tension current finds an alternative path to earth, instead of jumping the gap.

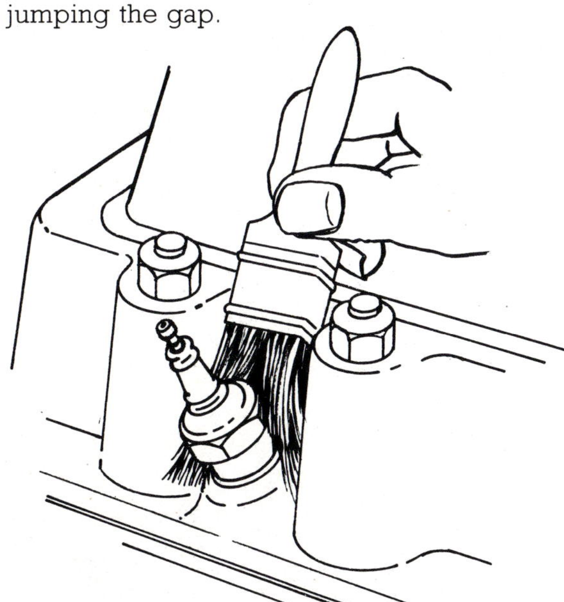

Most accessory shops sell sand-blaster type plug cleaners, usually powered by the cars battery - after use make sure all traces of sand are removed from the plug before it is put back into the engine. The 'sand' is in fact silicone carbide.

Check the gap between the central and earth electrodes, which should conform with the manufacturers specifications. If not reset it by bending the earth electrode - there is a special tool for this but if the gap is too small, the electrode can be bent using a small screwdriver and if too large it can be lightly tapped down with a small hammer or pair of pliers. Incidentally, even the gaps of new plugs should be checked and if necessary re-set before they are fitted.

For best results the tip of the central

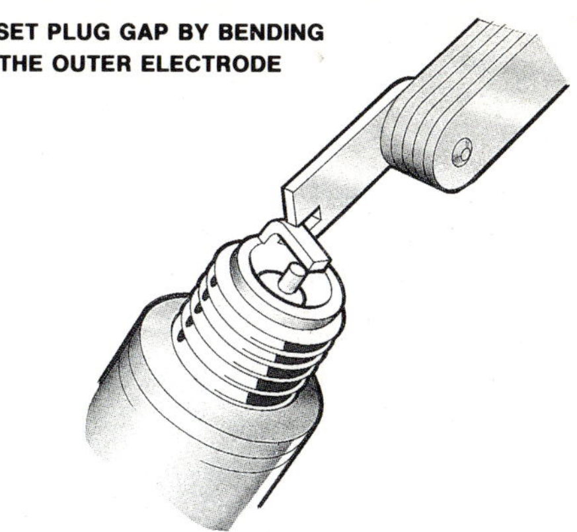

SET PLUG GAP BY BENDING THE OUTER ELECTRODE

electrode should be more or less flat. If due to erosion it has been rounded off, it should be filed flat again before setting the gap.

Sparking plugs are available with either flat seats, in which case a washer is permanently attached to the plug or with tapered seats, which do not need a washer. With either though it's important that the plug is fully seated, for if it isn't the heat dissipating qualities will be affected, in addition it is possible that some compression may be lost, allowing fuel/air mixture to escape into the engine compartment.

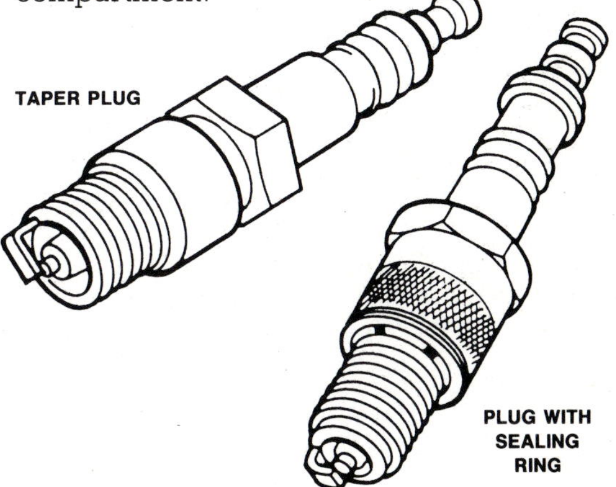

At the other extreme it's important not to overtighten the plugs, especially those with a tapered seat or they will be extremely difficult, if not impossible, to remove - in general and where no specific torque figure is given, tighten these down hand tight. then turn them through another one-sixteenth of a turn only.

The flat-seated variety (with a sealing washer) can be tightened through about one-quarter of a turn.

FAULT FINDING

Various surveys carried out by the motoring organisations over the last few years have consistently shown that more vehicles break down or refuse to start because of a fault in the ignition system, than for any other reason. The same surveys also showed that generally a large number of these ignition faults were entirely due to lack of maintenance.

However one hopeful sign with these statistics, is that they are not increasing in line with the car population as a whole. This is without question due to the fact that more and more manufacturers are fitting electronic ignition, which indicates that on the low-tension side of the ignition, it is the contact breaker which causes most trouble -either dirty and pitted or incorrectly gapped, perhaps both. Either way it results in a weak or non-existent spark, particularly on starting or in the higher speed ranges.

Dirt is also a factor in the most common type of problem in the high-tension circuit. Dirt coupled with moisture, will form an excellent alternative path for the high voltage current. If this should happen at the top of the coil or on the rotor arm, it would normally mean no spark whatsoever. However if it were just the distributor cap affected, the result could be the same or possibly a misfire on one or more cylinders, with some backfiring through the inlet manifold.

It is also possible for the HT current to leak down the side of a dirty and damp sparking plug, resulting in a misfire at that cylinder.

Most non-start situations on a cold and damp morning are entirely due to this combination of dirt and moisture and can, in some cases be rectified simply by spraying the system with one of the appropriate water displacing sprays such as WD 40 or Duckhams DPP. Once the engine has started, the build up of heat will also help in drying off the system.

In normal circumstances a well maintained and clean ignition system is unlikely to suffer from either of these problems. However this is no guarantee that nothing else could go wrong - it can, but by adopting a pretty well defined check pattern, ignition faults are generally easier to locate and rectify than those in other areas.

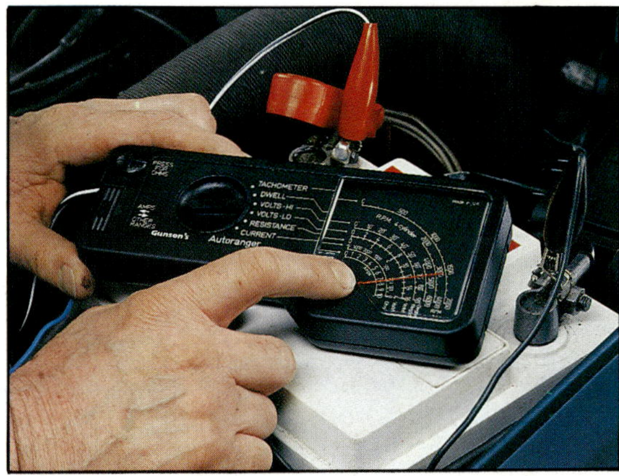

One method of testing the battery is to connect a voltmeter across its two terminals, then disconnect the ignition and spin the engine on the starter about ten seconds. Discounting any mechanical defect and severe cold conditions there should be no more than a two volt drop in the meter reading.

It doesn't necessarily follow that an engine failure or refusal to start is due to an ignition problem - even assuming that the battery and starter motor were functioning correctly, there could be something wrong with the fuel system or indeed, there could be a mechanical failure.

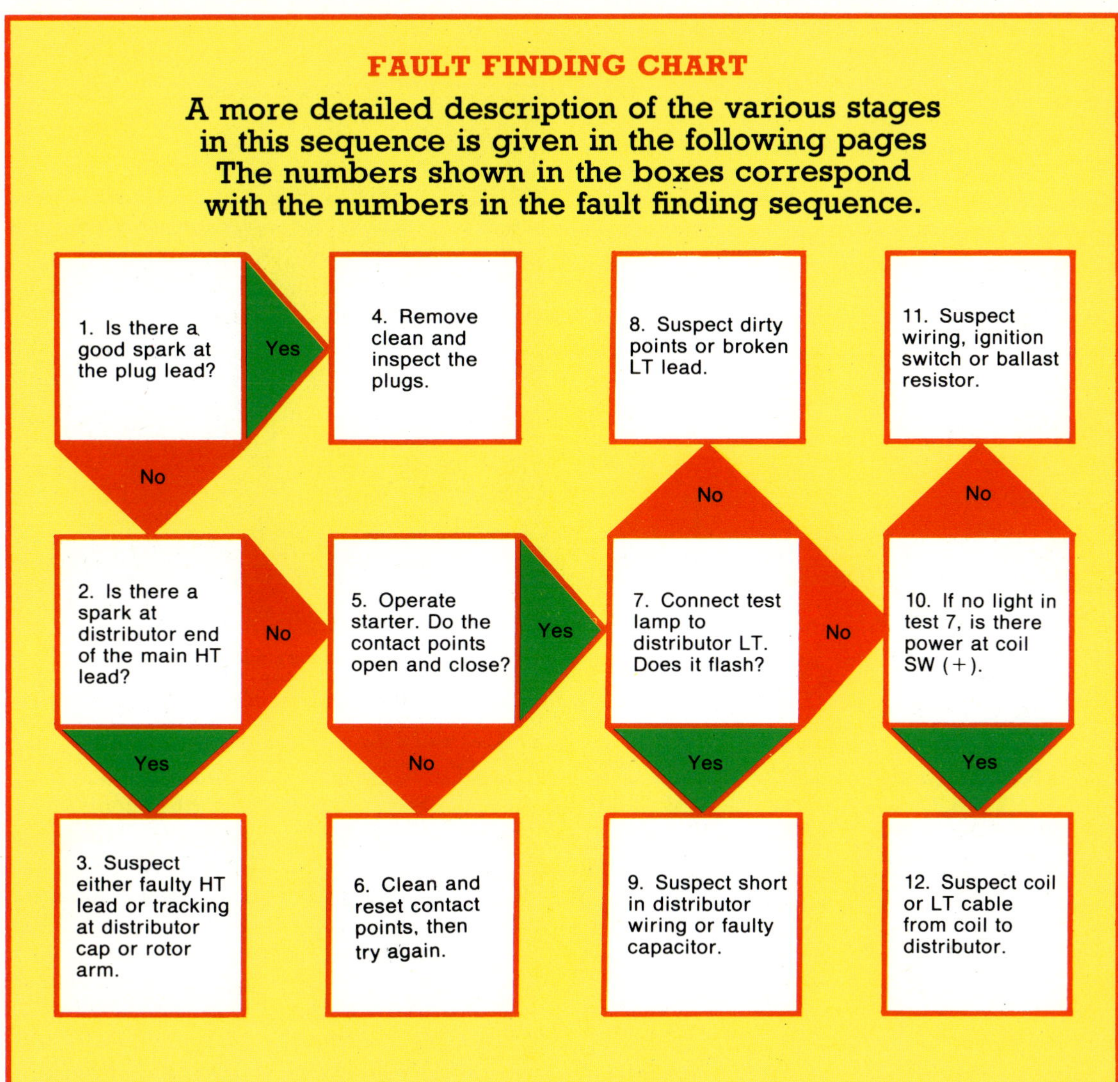

Operating the starter motor puts a heavy drain on the battery, especially when the engine and its oil are cold. If either the battery or starter motor is in poor condition, this drain could be such that there could be insufficient battery power left to adequately 'charge' the coil. A quick and simple check is to switch the headlamps on and turn the starter switch - if the starter motor operates and the lights dim slightly, both can be considered serviceable.

Severe dimming of the lights is indicative of the battery being in a poor state of charge, a high resistance at the main connections (including the earth return) or a defective starter motor.

However in general and unless the reason for an engine refusing to start is self evident, such as no fuel in the tank or a connecting rod through the side of the block, the first check in every case, should be for a spark at the plug - usually that at number one cylinder, but in reality any one would do.

The procedure from then on is as shown in the fault finding chart as shown above. Some of the checks call for the use of a test lamp, but in all cases a voltmeter can be used instead. However in those ignition systems using a ballast resistor the voltage readings at the coil terminals will be far below that of the battery, a test lamp will also glow far less brightly than normal.

FAULT FINDING SEQUENCE

1. This test involves removing a plug lead from the sparking plug and holding the end of it about 3/16in (5mm) away from a clear metal part of the engine, while an assistant operates the starter or alternatively use Gunson's Remotastart. If the ignition is in order a spark will jump from the lead to the engine.

Some plug leads may be fitted with recessed insulators, making it impossible to hold the end of the cable near the engine, in these cases insert a bolt, nail, screw or something similar in the end of the insulator.

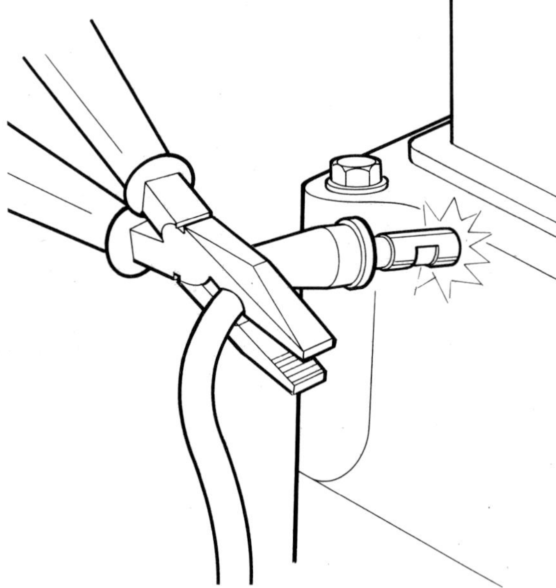

For obvious reasons do not hold the plug lead near the carburettor, the same applies to the rocker/cam cover for these could be retaining explosive gases.

2. If the first test resulted in a 'no spark' situation, the logical next test is to check for a spark at the distributor end of the main HT cable from the coil. To do this disconnect the lead at the distributor cap and then hold it as before, about 3/16in away from some metallic part of the engine, while an assistant operates the starter (the ignition must obviously be switched on, if separate from the starter switch). Other than on a single cylinder engine, the sparks produced in this test should be much more

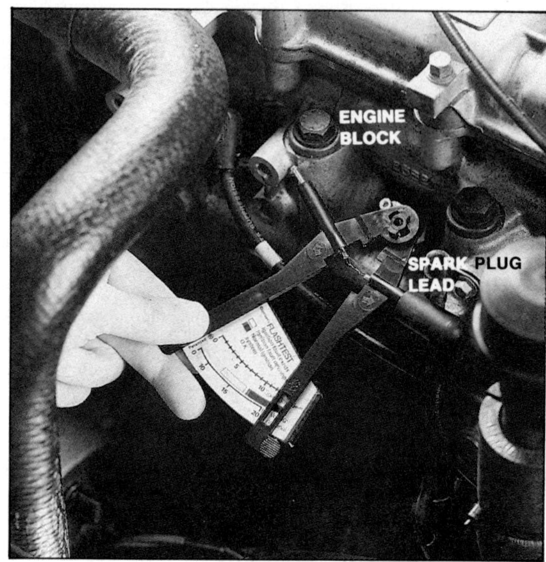

One of the ignition systems main functions is to produce a spark strong enough to ignite the mixture, it follows that by checking on the quality of the spark, all other checks (other than timing) can be disregarded. Gunson's Flashtest does just that. If the spark is good enough the system must be in order.

frequent than those from a plug lead.

If there is no assistant available, the same test can be carried out by first removing the distributor cap and then, with the ignition switched on, flicking the contact points open and closed. Every time the points open a spark should jump the gap between the end of the lead and the engine. Should the engine have stopped with the contact points open, simply bridge them with a screwdriver to produce the same result, as the screwdriver is withdrawn a spark should be produced.

3. Should the last test have produced a spark at the end of the main HT lead, but there still wasn't one at the plug, the fault could be in the HT lead to the plug (unlikely to happen at all the plugs on a multi-cylinder engine) or, more likely, tracking at the rotor arm or distributor cap.

It can sometimes be difficult to check a distributor cap for tracking, as one of the most likely areas where this occurs is around the main HT cable tower, which would have been disturbed in tests number 2. If tracking is thought to have occurred the cap should be removed and closely

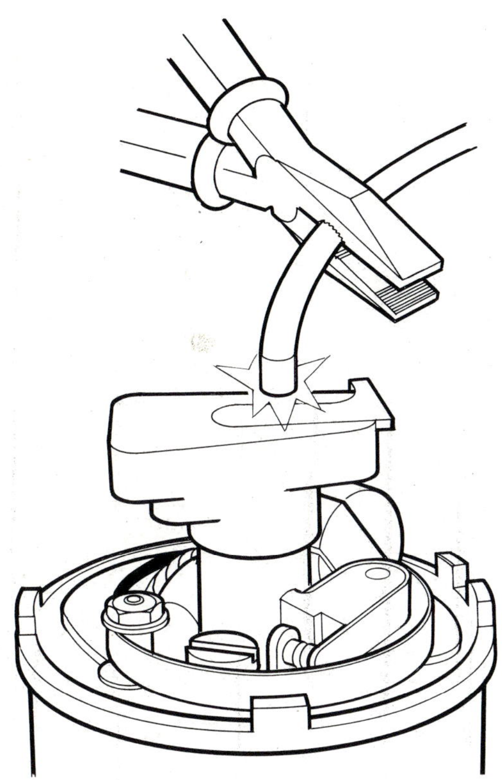

examined on both its internal and external surfaces, any apparent cracks or scratches emanating from a tower or segment should be treated as suspect.

The rotor arm can be checked for serviceability by carrying out a similar test to No. 2, but instead of holding the end of the cable near a metallic part of the engine, hold it near to the electrode of the rotor. If a spark jumps from the end of the cable to the electrode, then the rotor arm is faulty, for the electrical current is leaking to earth through the rotor. This can be due to dirt or moisture over the rotors insulated surface, in which case a simple clean and dry-off may clear the problem.

Most distributor caps incorporate a carbon brush and spring at the central terminal, this is used to transmit the HT current from the terminal to the rotor arm. If this has fallen out, jammed or disintegrated, the resulting large gap in the circuit from terminal to rotor could result in no current flowing and increase the likelihood of some tracking taking place. Should this be the case a short length of metallic foil, such as kitchen foil or a milk bottle top can be cut, rolled and fitted in place of the carbon brush as a temporary measure.

Should the lack of a spark at the plug be due to a faulty cable, or indeed if the main HT cable is defective, either could be replaced by a length of ordinary electric cable. This would need to be wired in such a way that it couldn't short out to anything metallic and would probably need supporting with string, alternatively it could be sheathed in plastic tubing, similar to that used in the windscreen washer system. Should there be no cable available an ordinary length of bare wire, suitably insulated, can be used instead.

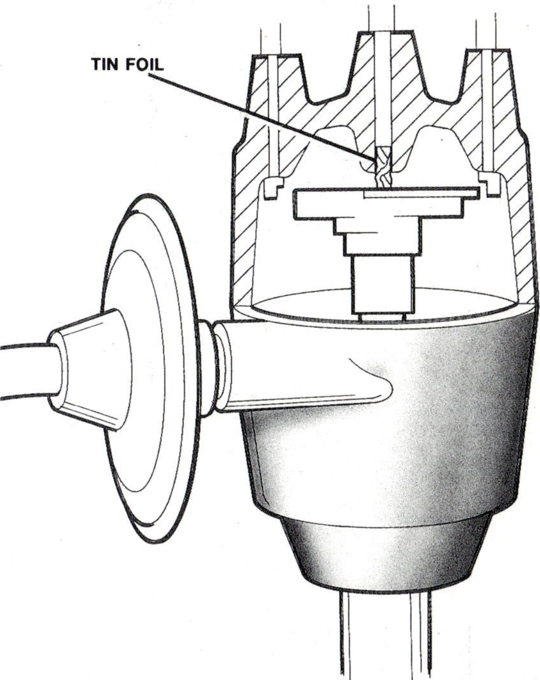

4. If the first test produced a good 'fat' spark at the plug lead, the next step should be to remove and inspect the sparking plugs. Should the plug electrodes be wet with petrol, it could be that the engine has been over-choked, although this could be coupled to an ignition or electrical problem which results in a poor spark, especially under cranking conditions. This situation is often made worse by the driver continually pumping the accelerator, so forcing even more fuel into the engine.

The immediate answer is to remove all of the plugs and dry them, preferably in an oven, and then try again. If the result is the

Sparking plugs can become oil fouled in an older engine

same, it is possible that the carburettor is delivering an excess of fuel. At the other extreme should the plugs be dry, it could be that the carburettor isn't delivering sufficient fuel - possibly a fault in the choke mechanism, although in these cases the engine will occasionally fire but not run.

If the plugs are found to be wet with water it's probably due to either a defective cylinder head gasket or a cracked cylinder head or block. This could result in serious engine damage if the leak was such that any cylinder became partially filled with coolant.

5. For this test the distributor cap and in most cases the rotor arm should be removed. Then with an assistant operating the starter, check that the points open and close. In addition, if possible check them for condition (see test number 8) - should they appear to be covered in a grey dust like material, they could be physically closing, but not making electrical contact. If this is so, try bridging them as in test number 2.

In some cases where access may be a problem, a small mirror would be a useful aid in carrying out this test.

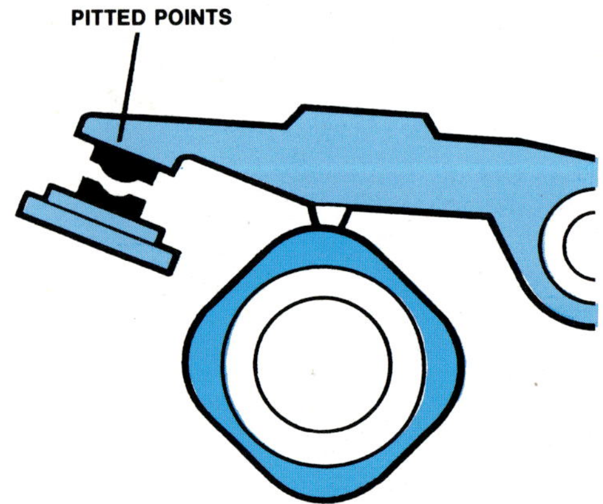

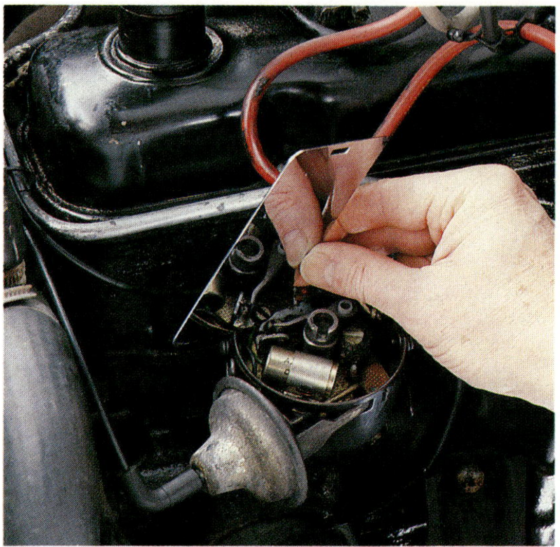

6. For maximum efficiency the condition and setting of the contact breaker is critical, for not only does it determine how well the coil is 'charged up' and therefore the quality of the spark, but if incorrectly set, will cause the spark to be produced at the wrong time (see page 28). The ignition timing should be checked after altering the contact point setting.

7. In this check a test lamp is connected between the low tension (LT) terminal at the distributor and earth. If there is no suitable attachment point it can be connected between the CB or - terminal at the coil and earth, in other words from either end of the thinner cable connecting the coil and distributor to earth. There is no need to disconnect the existing cables.

Now when the engine is cranked over with the ignition on, the lamp should normally go on and off as the points open and close. To prevent the engine from starting, disconnect the main high tension (HT) lead at the coil.

If there is a ballast resistor in the system, the light given off by the lamp will be of lower intensity than normal, a voltmeter should read in the region of 6 to 8 volts.

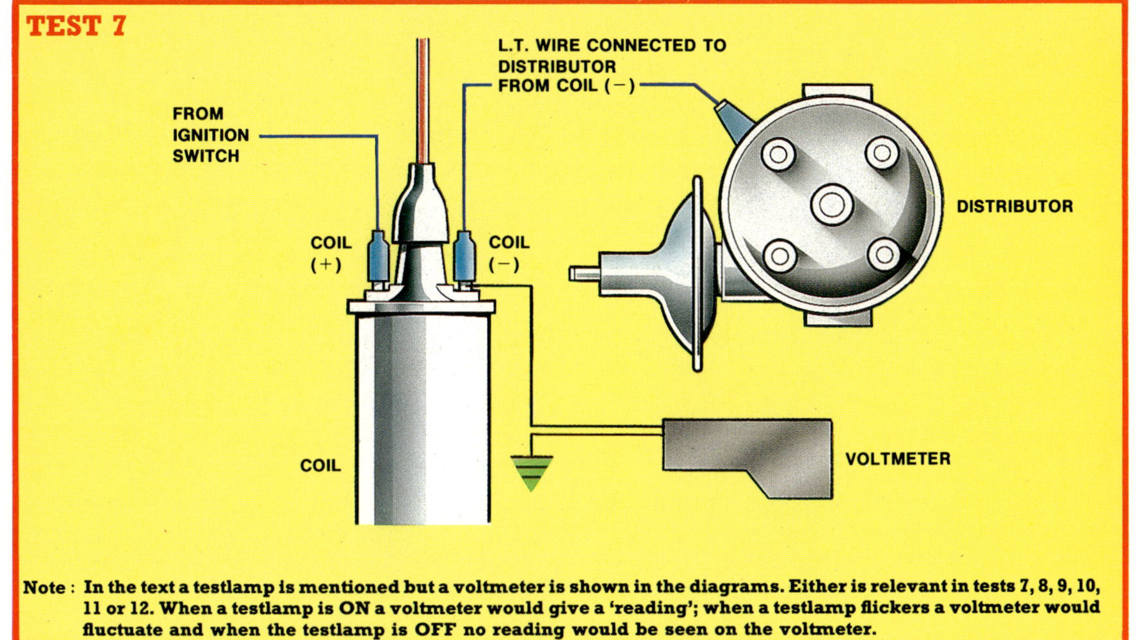

Note : In the text a testlamp is mentioned but a voltmeter is shown in the diagrams. Either is relevant in tests 7, 8, 9, 10, 11 or 12. When a testlamp is ON a voltmeter would give a 'reading'; when a testlamp flickers a voltmeter would fluctuate and when the testlamp is OFF no reading would be seen on the voltmeter.

8. Normally the lamp should go out when the points close, that is when the circuit through the points to earth is complete. Therefore if the lamp stays on there must be a break in the circuit. This could be that the points are not making electrical contact (see test number 5). Try bridging them with a screwdriver - if the lamp then goes out it is the contact points at fault.

However should the lamp remain on, it means there is a break in the circuit, possibly in the small feed cable to the contact breaker within the distributor. Alternatively if the lamp is connected to the coil CB terminal, the fault could be in the cable from coil to distributor.

9. If in the previous test the lamp flashed on and off in time with the contact points, but there was still no HT output from the coil (test number 2), suspect either the capacitor or the coil. However faults with either can also result in a low HT output and a poor spark. It is of course possible that there is an adequate output but it is 'tracking' to earth at the top of the coil.

The only practical method of testing a capacitor is by substitution.
However, as one of the reasons for its existence is to reduce arcing of the points, a check of these for serious burning should give some idea of the capacitors serviceability. In some cases the capacitor may develop an internal short, which would earth the LT current. If this is so the test lamp would not come on.

Although it is possible to check a coil using the ohmmeter mode of a multi-meter, once again the only practical method of testing it is by changing it for another. It has been known for coils to fail when hot but show no sign of any defect when cooled down again. This could for instance, result in an engine running perfectly normally for some time and then stopping, only to start again after an interval of about half an hour.

---

10. If in test number 7 the test lamp failed to come on at all, the first thing to check would be the lamp itself by connecting across the battery. If this proves to be in order suspect either a short to earth between the lamp connection and the contact breaker or an open circuit (break) further back in the circuit. To isolate one from the other, disconnect the LT cable at the distributor and wire the test lamp from the end of the disconnected lead to earth. If the lamp then lights up, it means there is a short somewhere between the distributor terminal and the contact breakers - either at the small feed cable itself or where it connects to the contact spring.

If the lamp failed to light up, first switch the ignition off and then re-connect it between the other LT terminal at the coil (often marked SW or +) and earth. Then switch the ignition on again.

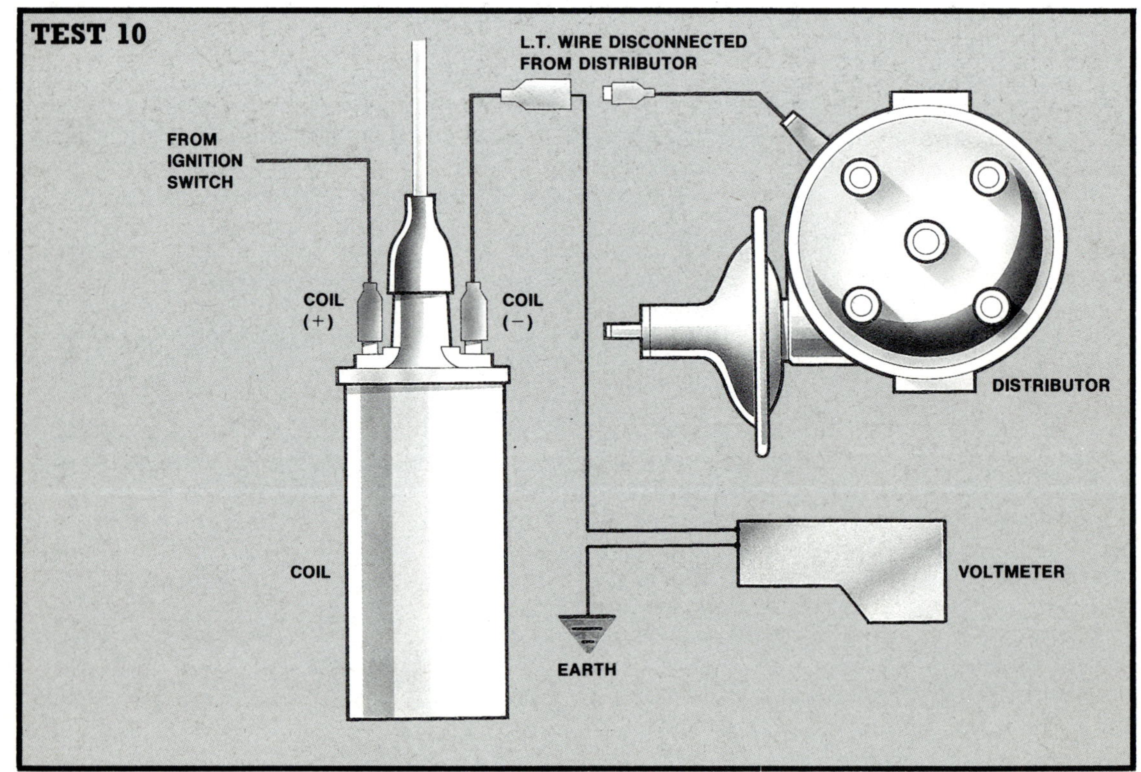

11. If the test lamp shows there is power at the coil SW terminal and none at the distributor LT connection (test 7), it indicates either a faulty coil or a defective lead between the coil and distributor. To isolate one from the other, check for power at the coil CB terminal. If the lamp still doesn't light up the fault is probably a break in the primary winding of the coil, whereas if the lamp does come on it indicates a faulty LT cable.

12. Should test number 10 show a lack of power at the coil SW terminal, it could be due to either a defective coil (primary shorting to earth) or a fault back in the circuit from the battery and ignition switch. To isolate one from the other disconnect the cable at the SW terminal and with the ignition switched on connect the

ELECTRONIC TEST METERS

Multimeters come in a variety of guises with different multiples of scales, or are quite small or very large, or light or cumbersome.

So first you must decide what your priorities of use are. Buy a well known make as these are likely to be of better quality, which means a higher standard of accuracy.

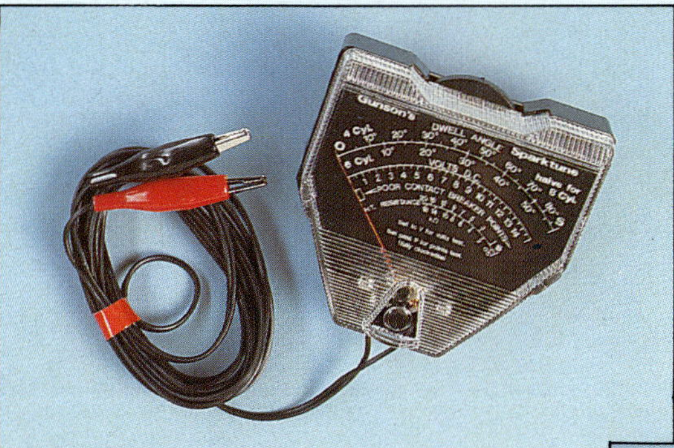

GUNSON'S SPARKTUNE

– is conveniently small but with several scales to measure dwell, volts, ohms and points condition.

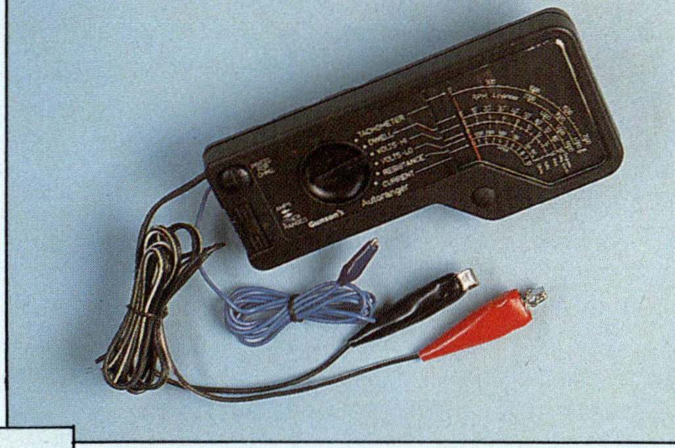

GUNSON'S AUTORANGER

– is more comprehensive with a tachometer and scales for volts, ohms, amps and points condition and is shaped for easy hand held use.

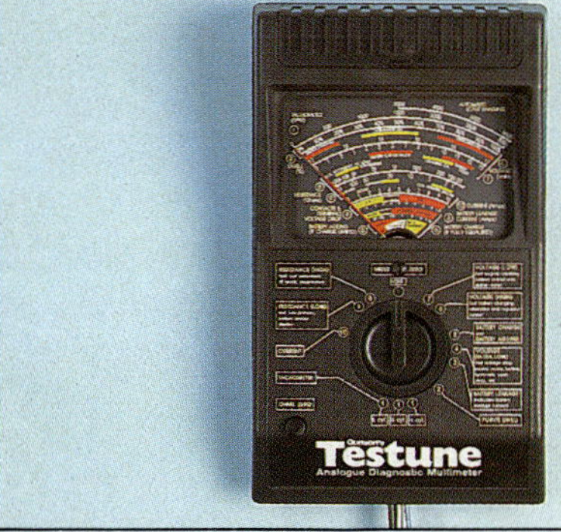

GUNSON'S TESTUNE

– is the most modern and very compact and thoroughly comprehensive meter for any vehicle test, presenting its information in a more understandable way than any previous one, while incorporating features found on no other multimeter.

test lamp between the end of the cable and earth. If the lamp lights up it means a defective coil or it could also be a short to earth in any radio interference suppressor that may be wired to the SW terminal - this can be easily checked out by disconnecting it and repeating the test.

No power at the SW terminal means a fault further back in the circuit such as a loose connection, a defective ballast resistor (if fitted) a faulty ignition switch or even, in some cases a blown fuse - if this is so try and discover why, before replacing it.

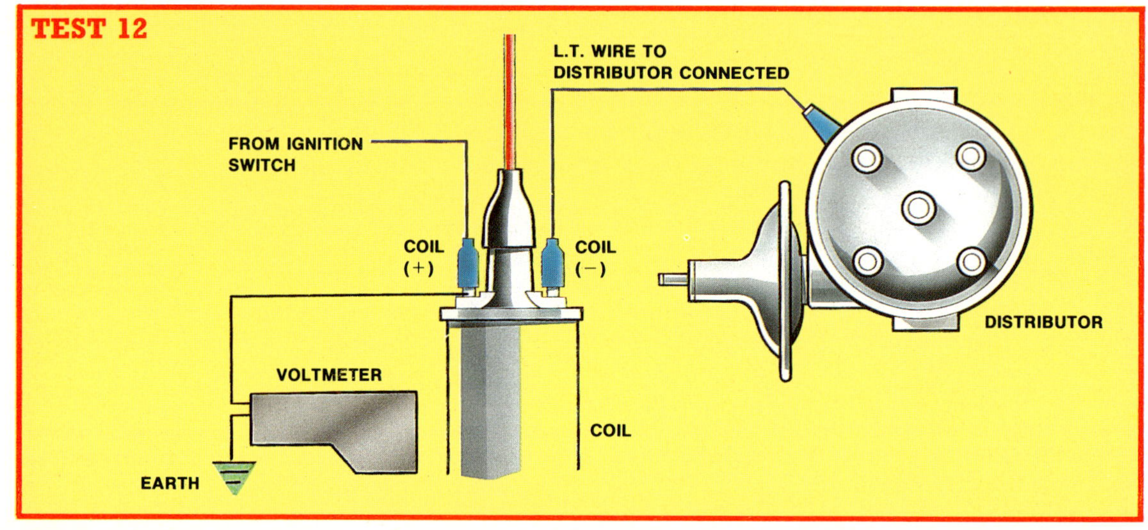

TRACKING (see test No 3)

Tracking is where the HT current finds its own path or track over an insulated surface (the distributor cap, coil tower, rotor arm or sparking plug). When this occurs the tracks can sometimes be seen as thin greyish-black lines. On the distributor cap for instance these lines may extend from the central terminal to one of the plug lead towers or segments, in which case the engine may only fire on one cylinder. In some cases however, the tracking may be between two segments causing a misfire. In both cases there could also be some backfiring through the inlet manifold.

If tracking is taking place on the coil tower or the outer surface of the distributor cap, it can sometimes be seen at night as thin streaks of lightning.

In extreme cases where the tracking follows a scratch mark or crack in either the cap or rotor arm, or where it has formed a definite path, it may be possible to clean the track out and fill it with a hard setting, non-conductive adhesive - most two pack epoxy resins would be suitable. Alternatively scratching a deeper track across the existing one and filling the new track with adhesive can form a barrier across the path of the leaking current. In some cases a hole drilled through the track would serve a similar role.

In an emergency a light smear of engine oil or a coating of nail varnish can be used instead of adhesive

COMBUSTION KNOCK

Combustion knock, detonation or as it is more commonly called 'pinking' are terms used to describe the spontaneous combustion of part of the fuel/air mixture in the combustion chamber, which causes the engine to give off a metallic tinkling sound, especially when accelerating or climbing a hill.

What happens is that when the plug fires the mixture, a flame front occurs and which spreads outwards from the plug. In a normal (non-Knocking) engine this flame front will progress steadily until it reaches the furthest part of the combustion chamber. When it does combustion is complete.

However the mixture in this furthest part of the combustion chamber, usually referred to as the 'end gas', is being both heated and compressed even more by the approaching flame front. It can, in fact self-ignite.

If this happens the piston will receive a sharp blow instead of a progressive push, causing the metallic tinkling noise.

Light knocking when accelerating is relatively harmless, but in a more persistent or heavy form it results in loss of power, in addition the excessive shock loadings on the piston crown and overheating that occurs can damage the piston.

Using lead in the fuel is one way of reducing the chances of this happening as the lead precipitates out in the form of a fine dust during the combustion process. The dust with its large surface area, acts as a combustion inhibiting agent, so reducing the likelihood of that end gas exploding.

This phenomenon shouldn't be confused with pre-detonation or pre-ignition, which is when the charge is ignited by some hot-spot in the combustion chamber, before the plug fires. Although the symptoms (noise) may be similar, this is potentially the more dangerous problem.

Pinking can be caused by an engine being out of tune, a build up of carbon deposits in the combustion chamber or by using a low grade petrol and has been more of a problem in the UK since the reduction of lead levels in fuel in January 1986. If a tune-up or change in the grade (and brand) of petrol doesn't have any effect, de-carbonising the cylinder head (de-coke) may help – this can sometimes be averted by adding an upper cylinder lubricant (Redex or similar) to the petrol, although do not expect instant results, using this method.

If all else fails retard the ignition by 3 or 4 degrees.

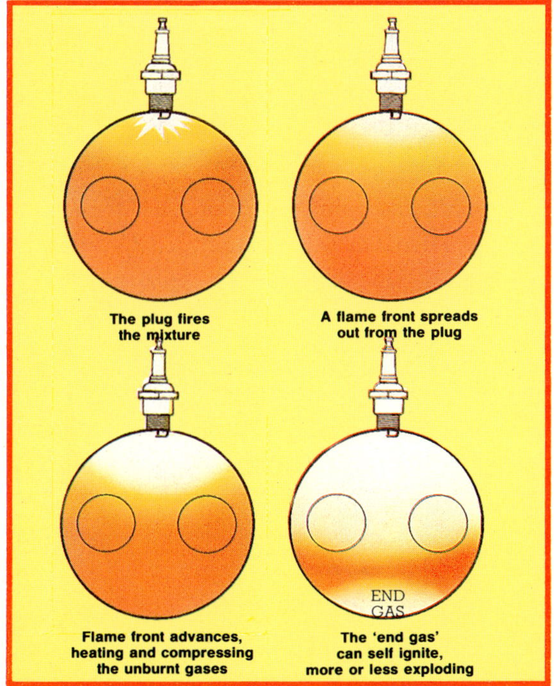

The plug fires the mixture

A flame front spreads out from the plug

Flame front advances, heating and compressing the unburnt gases

The 'end gas' can self ignite, more or less exploding

RUNNING ON

Sometimes referred to as dieselling, this is a condition where the engine continues to run, albeit somewhat 'lumpy' after being switched off.

Much like pre-detonation it is caused by some hot-spot in the combustion chamber igniting the fuel/air mixture. This can be carbon deposits, incorrect grade sparking plug, general overheating, poor design (sharp edges in the combustion chamber glowing incandescent red) or an engine out of tune.

GLOSSARY OF TERMS

Ampere	Electrical unit of measurement used to record the flow of electricity in a circuit. Generally referred to as amp or amps and measured with an ammeter.
Automatic advance	A term used to describe process of making the spark occur earlier (advanced) as the engine speed increases. See centrifugal advance.
Ballast resistor	Special wire or resistor which reduces the voltage applied to the coil in the ignition circuit. It is by-passed when the starter motor is in operation so as to provide a stronger spark for starting purposes.
Brush, distributor cap	A small rod of carbon in the centre terminal housing of the distributor cap which is spring loaded into rubbing contact with the rotor arm.
Camshaft	The second shaft of the engine which rotates at half crankshaft speed and is used to control the opening and closing of the valves. In most engines it also drives the distributor.
Capacitor	Sometimes called a condenser, this is a non-conducting component that can store an electrical charge and release it when required to do so.
Capacitor discharge ignition	An electronic ignition system in which a capacitor is charged up and then made to discharge through a transformer, to produce a very high energy but short-lived spark at the plug.
Centrifugal advance	A method of advancing the spark, using of springs and pivoted weights in the distributor.
Circuit	In electrical terms a path or track along which electricity can flow.
Circuit diagram	An illustration showing the layout of a number of circuits.
Coil	As used in the ignition is made up of primary and secondary windings around a soft iron core and transforms the voltage supplied by the battery to that required by igniting the mixture.
Contact breaker	A mechanically operated switch in the distributor which is used to interrupt the primary current.
Contactless ignition	A version of electronic ignition in which the system is triggered by means other than the contact breaker.
Conventional ignition	A term often used to describe ignition systems where the coil primary current is switched by the contact breakers.

Detonation	A condition in the combustion chamber where part of the fuel/air mixture self detonates, causing a shock wave which produces a metallic tinkling sound known as pinking. Among other causes this can result from over-advanced ignition.
Digital ignition	A form of electronic ignition, where the timing of the spark is controlled by a microcomputer. Ignition data is stored in the computers in the form of three dimensional maps, providing the optimum timing for all engine conditions.
Distributor	A device used for directing the high tension current to the sparking plugs. In most applications it also houses the contact breaker capacitor and centrifugal advance mechanism.
Dwell meter	A meter used to measure the dwell angle. In most cases it is one mode of a multi-meter instrument.
Dwell period	The period of distributor cam movement, when the contact points are closed. This can be quoted in degrees and termed the dwell angle, or as a percentage and termed the dwell period.
Dynamic timing	A phrase used to describe the method of checking the ignition timing with the engine running.
Electronic ignition	Ignition systems employing electronics to switch the coil primary current.
Feeler gauge	A thin strip of metal of a pre-determined thickness used for measuring purposes (contact-breaker). Non-metallic (plastic) gauges must be used for measuring the air gap in some electronic ignition systems.
Firing order	The order in which the sparking plugs fire the mixture in a multi-cylinder engine. A typical in-line four cylinder firing order would be 1,3,4,2.
Four stroke principle	A cycle of operations in an engine whereby the piston moves through four stroke to complete the cycle.
Hall effect	Triggering device used to initiate the ignition process. Named after the American E. H. Hall.
High tension	Term used to describe that part of the ignition system which operates at very high voltages.
Generator	Machine used to generate electricity. A DC generator is termed a dynamo and machine that (initially) produces an AC current is called an alternator.
Ignition timing	The time at which the spark occurs. If this should be late the ignition is said to be retarded, whereas advanced ignition means that the spark occurred earlier than it should.
Low tension	Term used to describe that part of the ignition system which operates at battery voltage or less.

Map ignition	A term sometimes used for digital ignition systems.
Multi-meter	An instrument capable of making a number of electrical measurements
Ohm	Electrical unit of resistance. Measured with an ohm-meter.
Ohms law	Basic law of electrical theory which shows the relationship of electrical pressure (volts) resistance (ohms) and current flow (amps).
Pinking	The common name for the metallic tinkling noise produced by either detonation or pre-ignition.
Pre-ignition	Also known as pre-detonation, this is a condition where the fuel/air mixture in the combustion chamber is ignited before the spark occurs. Can produce similar symptoms as detonation.
Primary winding	The winding in the coil which carries battery voltage or less.
Rotor arm	A non-conductive rotor mounted on the distributor shaft with a metal electrode on its uppermost surface, which provides a path for the high tension current from the centre of the distributor cap to segments around its outer circumference.
Running-on	Term used to describe the tendency for an engine to keep on firing after being switched off.
TAC ignition	Short for Transistor Assisted Contact, this is a form of electronic ignition system which uses the contact breaker points to initiate the ignition process.
Secondary winding	The winding in the coil which carries battery voltage.
Timing mark	Mark or marks on either the flywheel or crankshaft pulley which are used in conjunction with marks or pointers on the engine to time the ignition.
Tracking	Term used to describe the phenomenon where HT current finds its own path or track over an insulated surface.
Transistor	A solid state electronic device which can be made to function as a remote controlled switch in much the same way as an electrical relay.
Two stroke principle	A cycle of operation where the piston moves through two strokes to complete the cycle.
Vacuum Advance	A method of making the spark occur earlier when the engine is lightly loaded.
Volt	Electrical unit of pressure, sometimes referred to as potential difference or electro-motive force (emf). Measure with a voltmeter.

INDEX

Produced by EMBS Ltd., Canvey Island, England

Art Director: Robert Onasse
Series Editor: Keith Hayward
Designer: Martin Salmon
Illustrators: Kevin Slater, Ronnie Capp
Artists: Kevin Slater, Alan Foster
Photography: Photocall Professional, Enfield, England
Typeset and Printed by Valentine Press Ltd., Laindon, England.

The contents of this book are believed correct at the time of printing. Nevertheless, the publisher cannot accept any responsibility for errors or omissions, or for changes in details given.

All rights reserved. No part of this publication may be reproduced, stored in a retrieval system, or transmitted in any form or by any means – electronic, mechanical, photocopying, recording or otherwise, unless the permission of the publisher has been given beforehand.

Published by Gunson Ltd, London, England

The Author and Publisher wish to acknowledge their gratitude to the following companies.
Automobile Association
Autocar Electrical Equipment Co. Ltd
Bosch Ltd
Champion Sparking Plug Co. Ltd
Lucas Electrical Ltd
NGK Sparking Plugs (UK) Ltd
Rover Group plc